SOIL MICROBIOLOGY

SOIL
MICROBIOLOGY

R. R. MISHRA
Professor
Botany Department
North Eastern Hill University
Shillong – India

CBS

CBS Publishers & Distributors Pvt. Ltd.

New Delhi • Bengaluru • Chennai • Kochi • Kolkata • Mumbai
Hyderabad • Uttarakhand • Nagpur • Patna • Pune • Jharkhand

ISBN: 81-239-0455-X

First Edition: 1996
Reprint: 2000, 2003, 2010, 2014, 2020

Published by **Satish Kumar Jain** and produced by **Varun Jain** for
CBS Publishers & Distributors Pvt. Ltd.,
4819/XI Prahlad Street, 24 Ansari Road, Daryaganj, New Delhi - 110002
delhi@cbspd.com, cbspubs@airtelmail.in • www.cbspd.com
Ph.: 23289259, 23266861, 23266867 • Fax: 011-23243014

Corporate Office: 204 FIE, Industrial Area, Patparganj, Delhi - 110 092
Ph: 49344934 • Fax: 011-49344935
E-mail: publishing@cbspd.com • publicity@cbspd.com

Branches:
• *Bengaluru:* 2975, 17th Cross, K.R. Road, Bansankari 2nd Stage,
  Bengaluru - 70 • Ph: +91-80-26771678/79 • Fax: +91-80-26771680
  E-mail: cbsbng@gmail.com, bangalore@cbspd.com
• *Chennai:* No. 7, Subbaraya Street, Shenoy Nagar, Chennai - 600030
  Ph: +91-44-26681266, 26680620 • Fax: +91-44-42032115
  E-mail: chennai@cbspd.com
• *Kochi:* Ashana House, 39/1904, A.M. Thomas Road, Valanjambalam,
  Ernakulum, Kochi • Ph: +91-484-4059061-65
  Fax: +91-484-4059065 • E-mail: cochin@cbspd.com
• *Kolkata:* 6-B, Ground Floor, Rameshwar Shaw Road, Kolkata - 700014
  Ph: +91-33-22891126/7/8 • E-mail: kolkata@cbspd.com
• *Mumbai:* 83-C, Dr. E. Moses Road, Worli, Mumbai - 400018
  Ph: +91-9833017933, 022-24902340/41 • E-mail: mumbai@cbspd.com

Representatives:

• Hyderabad: 0-9885175004	• Nagpur: 0-9021734563
• Patna: 0-9334159340	• Pune: 0-9623451994
• Jharkhand: 0-9811541605	• Uttarakhand: 0-9716462459

Printed at:
J.S. Offset Printers, Delhi (India)

PREFACE

An attempt has been made in this book to expose the students of soil microbiology to certain new ideas and facts available in the field. The author has tried his best to see the syllabi of various Universities both of undergraduate and postgraduate and he feels the information presented in the book will satisfy the need of the people interested in the microbiology of soil - a very fascinating area. All effort has been made to compile the recent informations, however, perfection is never achieved and as such any suggestion for further improvement by the readers will be welcomed.

The author has always been inspired by a team of his students who have been helpful in many ways in the preparation of the book and all my thanks to every one of them.

My family members have also helped in various ways and they deserved all my appreciations.

R. R. MISHRA

SHILLONG

CONTENTS

CHAPTER 4
MICROBIAL DECOMPOSITION OF HERBICIDES

2,4-D, 2,4-trichlorophenoxy acetic acid, phenyl carbamate, phenyl urea, acylanilide, malathion etc.

CHAPTER 5
TRANSFORMATION OF MINERALS

Transformation of phosphorus, sulphur, iron, manganese.

CHAPTER 6
NITROGEN CYCLE

Introduction, nitrogen cycle and microorganisms, nitrification, nitrate reduction, nitrogen fixation - non symbiotic nitrogen fixation, bacteria, blue green algae; symbiotic nitrogen fixation - legumes and *Rhizobium*, nodule formation, biochemistry of nitrogen fixation - actinorrhizal system, leaf nodule associations, *Azospirillum* species, factors affecting nitrogen fixation.

REFERENCES

Chapter 1

Soil structure and components

In order to understand the microbial population and its activity in soil, the structure of soil needs proper understanding. The physical and chemical properties of soil have a dominant role to play on the soil microorganisms. The outer loose material of earth surface, which is distinct from the underlying bedrock is called soil. It is from the outer layer that plants derive nutrients for their survival and also obtain their mechanical support. This outer layer, so called soil, is rich in organic substances and on account of a multitude of organic substances in the region, it embodies a vast population of living organisms. Bacteria, actinomycetes, fungi, algae and protozoa form the major constituents of the living population of soil.

Soil as such is composed of five major components - mineral matter, water, air, organic matter and a living population. These components vary from place to place and sometimes variation is noticed even within the same locality at different periods of time. To a great extent, however, the mineral and organic matter for a locality is relatively fixed. On the other hand, water, air and living population are always in a dynamic state and any change in the soil environment has a quick effect on the living population of the soil. Inorganic constituents of soil, because of their influence on nutrient availability, water retention and air, exert a major influence upon the living population. Size wise, the different particles of the mineral fraction vary from those visible to the naked eyes, to smaller microscopic particles called clay particles. Larger particles like stone and gravel exceed the diameter of 2.0 mm and the size of smaller sand particles ranges between 0.05 to 2.0 mm. Still smaller units, classified as still have a diameter between 0.002 to 0.05 mm and those with a diameter less than 0.002 mm are categorised as clay particles. The chemical properties and activities of the particles are directly related to their surface area. The smaller the particles, the more is the surface area exposed for various activities. On account of this, clay particles are considered to assume more significance in relation to the living population. This is mainly because of the greater surface area exposed per unit mass of the clay particles. The clay minerals are rich in silicon, oxygen and aluminium and they also contain iron, magnesium, potassium, calcium, sodium and other elements. The relative amount of each element in the clay varies.

The biological properties of soil are governed to a great extent by the texture of the soil. This in turn is determined by the content of sand, silt and clay.

It is also interesting to study soil depthwise. A vertical section of the soil reveals a distinct profile. Each profile is characterised by several horizontal layers which are known as horizons. The structure, texture and colour of each layer is distinctly different. Normally each profile is composed of three major layers which are designated as A, B and C horizons. Each horizon is further subdivided as A_1, A_2 and A_3 etc. Usually a typical profile may exhibit four zones.

(a) The upper zone of decaying organic debris, -A_0 horizon.
(b) The layer next to the upper zone is poor in certain inorganic constituents because they normally get removed during soil formation.
(c) The next horizon lies at a greater depth, and in this are deposited some of the constituents from the upper layers.
(d) The bottom layer is very similar to the original material from which the soil has developed.

Horizon A is of the greatest biological concern because roots, small animals and micro-organisms are profoundly available in this region. Besides these, the concentration of organic matter is also maximum in this region which provides nutrients to the microbial population. In horizon B, the organic matter is less and so too plants roots, which in turn leads to the low microbial population. True soil is mainly composed of horizons A+B. In the C horizon, organic matter is very low and it is mainly composed of the parent material of the soil profile. This layer, on account of these characteristics, shows very little microbial life. Relative populations of the different mineral constituents like gravel, coarse and fine sand silt and clay fractions determine the texture of the soil. In relation to fertility of the soil, its mineral aspect is also of great importance. Generally, in weathered and leached soil, the amount of quartz grains may vary from 90-95% of the sand fraction, however, the quality of the grain in new young soil is approximately 60% only. In such a case the remaining part is mainly composed of plaguiolase, angite and olivine (Leeper, 1964). This fraction is very important in soil, and this on weathering, provides a continuous supply of the elements useful for the growth of the plants.

Clay mineral part of the soil is equally important and it accounts for the physical properties of the soil and to a great extent, the base exchange capacity is determined by the amount and the nature of the clay minerals. Sometimes, a single clay mineral may be responsible for the clay fraction of the soil, however, in most of the cases several minerals join hands in the process. Besides amorphous organic matter, negative charges on the surface of the clay particles are responsible for the cation exchange capacity of a soil.

THE ORGANIC MATTER

The natural organic substances, which sooner or later are added to the soil, contribute significantly to the total soil organic matter. Depending upon the

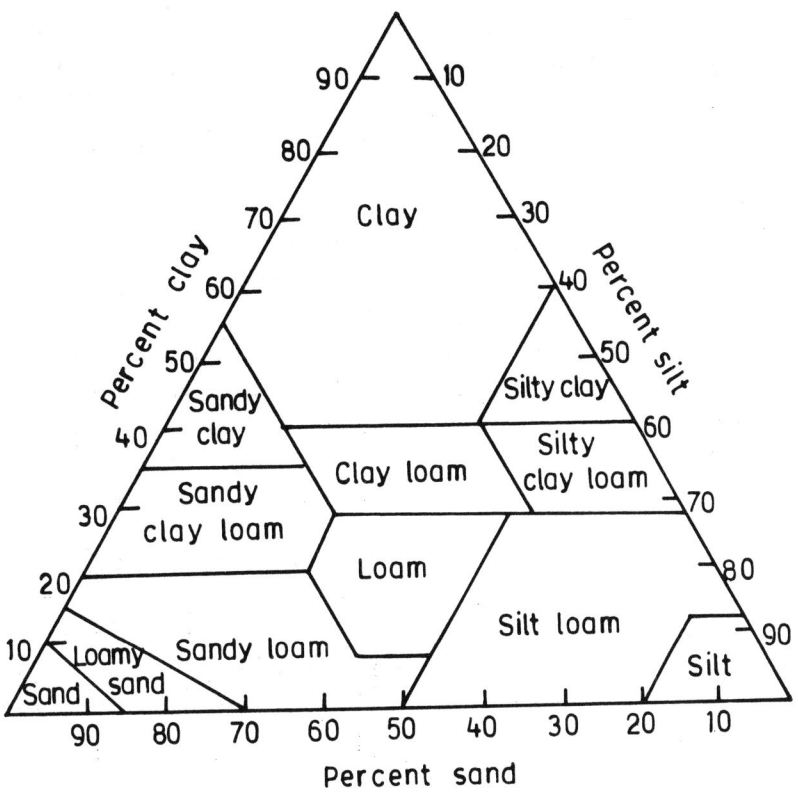

Fig. 1.1(a) Percentage different soil Constituents.

chemical nature of the organic substances, their stay in the soil may be brief or relatively long. Chemically simple substances are readily decomposed by the soil microorganisms and they soon contribute to the organic fraction of the soil. On the other hand complex substances take time for their decomposition, and thus the process of formation of organic matter in such situations, is naturally slow. Much of the organic matter, which sooner or later is incorporated into the soil is either of plant or animal origin, which gets accumulated on the surface of the soil in the form of debris. The debris once accumulated on the surface of the soil may have different fates.

Alternatively, through the action of earthworms and other animals, the debris may be incorporated into the soil much earlier and in such conditions, decay is achieved when the organic matter is incorporated in the mineral soil.

Considerable studies are available for the microbial decay of organic substances. By and large, information in relation to plant decay is more available. Water soluble materials like starches and the proteins disappear during early stages of decay. Relatively more complex substances like hemicellulose and cellulose follow next for decomposition. Subsequently, residues left behind are rich in lignin and cuticularised cell wall and they are the last in the chain of the decomposition process. Similarly, the animal remains on the soil surface also follow a certain sequence in the decomposition process. Soft protienaceous materials of the smaller animals decay rapidly leaving chitinous skeletons.

The total amount of organic matter in soil varies considerably. The maximum 10 to 12% dry weight of organic matter is reported in good grasslands Chernozem. In disturbed situations like a cultivated field, the amount may be as less as 1%. Humic acids embrace the substantial fraction of the organic matter and they account for 80 or 90% of the total amorphous organic material. Other substances like amino acids are present in very small quantity. They account for about 0.1% of the total organic material. Humic acids contain a certain amount of nitrogen. It was also suggested that humic acids are complexes of protein and lignin residues. The amount of nitrogen in humic acid, however, varies for the soil of different environments. It has been assessed that the humic acids are mostly aromatic in constitution and approximately 30% of material is aromatic (Burges, Hurst and Walkden, 1964). Normally lignin present in the soil affects the nature of the humic acids produced. It has been noticed that the phenolic fractions are released during the decomposition processes in the soil. Carbohydrates like glucose appear in much less quantity. Similarly, starches and pectins are also available in small amounts. Polysaccharide, however, mostly hemicelluloses are present in a relatively significant amount.

This release is either from the decomposing lignin or the flavonoid residues, or from the metabolic activities of microorganisms. The phenolic fractions are then polymerized. The proportions, however, may vary from place to place and also perhaps from time to time.

The soil moisture

Croney (1952) categorised soil moisture into the following three categories.

(a) Gravitational water - which moves through the soil under the influence of gravity,

(b) ground water - this is held below the water table, and

(c) held water which is retained in the soil after gravitational movement has ceased.

In reasonably good drainage conditions, the amount of water in soil varies. Weather conditions play an important role in the soil water regime. In temporarily flooded conditions and during heavy rain, the pores between the soil particles are filled with water which once the water has drained off, are again occupied by soil atmosphere. Soon after the drainage of the gravitational water, a relatively stable condition in the soil moisture regime is attained. This stability is on account of the balance of the gravitational pull by the capillary and the adsorption forces exerted by the interstices and the colloidal materials in the soil. The moisture content of the soil under such a situation is called the field capacity and this is normally used to measure the moisture level of soil. In drier conditions, when the moisture content of the soil falls below the field capacity, extraction of the water from the soil by plant roots becomes difficult. If the situation deteriorates further, the plants can no longer absorb water from the soil and it leads to their permanent wilting. This is designated as permanent wilting point of a soil and interestingly enough this value is more or less constant with little variation. There is a lot of difference between the moisture content of a soil expressed on the basis of dry weight and the permanent wilting point for different soils. Thus, in order to have a clear picture of the water retention capacity of a soil for the purpose of plants - it is more appropriate to have a clear understanding of how much water is really held in soil.

Schofield (1935) coined the PF scale to meet this need. This scale expresses on a logarithmic basis the difference between the free energy of the moisture in the soil system and a free water surface. Using this scale a soil at field capacity has a PF of 2 and permanent wilting point is approximately PF 4.2; oven-dry soil on the other hand exhibits PF-7.

The Soil atmosphere

Several processes operating simultaneously in the soil are responsible for the soil atmosphere which is usually in a very dynamic state. The living population of the soil converts the available O_2 of the soil into CO_2 and the fresh O_2 subsequently diffuses down from the soil surface. CO_2 on the other hand diffuses from the soil into the atmosphere. The relative proportion of the two gases i.e. O_2 and CO_2, will depend upon the rate of production of CO_2 and on the ease of diffusion. Normally in drier soil the level of CO_2 seldom rises above 1%. However, in water-logged conditions when the pores get filled with liquid, diffusion is affected and the level of CO_2 rises, which ultimately leads to a fall in the level of O_2. It has been observed that with the onset of rains, there is a sudden rise in the microbial activity and consequently an increase in the level of CO_2. In many cases the rise in CO_2 level is 3 or 4%, or in exceptional cases it may rise upto 10%, though such situations are relatively of short duration. Correspondingly there is also a drastic fall in O_2 levels and sometimes, in water logged condition, the O_2 level may fall down to zero.

Normally in litter layers the CO_2 level never rises above 0.5% and in such situations the microbial life in the soil is not affected. However, when the CO_2 level rises to 3 or 4% the life of microbes in the soil, particularly of fungi, is greatly affected adversely. Burges and Fenton (1953) observed that the fungi at deeper layers of the soil are more resistant to higher levels of CO_2 than those present at the upper layers. It is however difficult to explain such adaptations of fungi at the two layers. Interestingly enough, this behaviour of the soil fungi has wide implications. Garrett (1956) suggested that local concentrations of CO_2 in the soil are quite important in determining the activity of some parasitic fungi. The soils rich in CO_2 and acidic in nature are less prone to root pathogenic fungi than those well aerated ones with an alkaline nature.

Living Population

As indicated earlier, the living population of the soil is composed of five major groups; bacteria, actinomycetes, fungi, algae and protozoa. Earlier workers were of the opinion that by and large bacteria are the only major constituents of soil microbial life and infact soil microbiology was then considered to be just aliens and of not much significance. This contention continued for a long time but later on it became abundantly clear that such a concept does not have much relevance, and besides bacteria, the other four groups of organisms also have equal importance, and then they were also grouped as a major constituent. However, it is a fact that bacteria are the most abundant and they are more numerous than the other four groups of organisms combined together.

Minor fluctuations in the O_2 level of soil do not normally affect the soil microorganisms. Only in exceptional cases when the O_2 level falls below 5% does the microbial activity of the soil gets adversely affected.

Bacterial cells are unique in some aspects and they exhibit extreme variability. Infact, there was a time when the theory of plomorphism was strongly advocated. This concept was based purely on the variable nature of bacteria. However, this concept is not accepted now and it has been convincingly proved that bacterial species are stable like any other plant and animal. It is almost proved now that bacteria do not change from cocci to bacilli or mutate from one genus to other. However, within the species a great variation in biological activity, antigenicity and virulence exists. To explain the variability of fungi three following theories have been proposed :

(i)   Adaptation or lasting modification

(ii) life cycles, and

(iii) utation and selection.

Lately mutation and selection have been considered to be explanations for variation.

Like in other forms, in bacteria also, the mutation is a spontaneous, unidirectional change and it results in heritable change. On mutation, a mutant results in a mutant clone, which becomes apparent in the form of an inherited phenotypic change in morphology or physiology of the cell. Subsequently the new variants normally breed true and they retain the new inherited property. There does, however, exist the possibility, that even the new mutants may undergo mutation and may revert back to the parent life. It is easy to make morphological changes in the mutants, but the physiological changes are difficult to ascertain, and it can be done by use of a selective environment. Replica (Lederbeg, 1952) plating method has also been used for such detection. Spontaneous mutation are the resultant of an accidental error in the copying of a gene at the time of reduplication. It has been noted by Szilard (1950,51) in *E. coli*, that the mutation rate is independent of the generation time and it is a result of metabolic activity. On the other hand, Within (1953) and Zamenoff (1945) found that a definite correlation exists between mutation rate and cell division in bacteria. Despite of all these controversies, it is presumed that mutation is a genetically controlled phenomenon.

Types of mutants

As a consequence of genetic changes, bacterial cells exhibit a variety of mutants, which are manifested in a number of ways. Marked colonial and morphological changes, differences in nutritional requirements, changes in virulence and antigenic characteristics are some of the obvious manifestations. We know clearly that a single gene may control the formation of a single enzyme, as a consequence of it any change in a particular gene will be reflected in the functioning of the related enzyme, and thus the specific biochemical function of the enzyme will be affected. Ordinarily through mutation the chemical components of the bacterial cells not essential for the existence of the organisms are affected. Sometimes, however, effects may be such that even the existence of a particular organism is endangered, and then the mutation results in being a lethal one.

Systematic study of bacteria

Exact identification on account of variability is a difficult task. Many a time the classification of bacteria on the basis of their morphology and arrangement of cells is not possible. As such for classification of bacteria, besides morphology, other characteristics like the cultural and physiological ones are also taken into consideration. Shape and size of the cell, in case of the cylindrical forms, the shape of the ends of the rods; the spore of the cell and the nature of the flagella; in case of the spore forming species, their ability to produce spores and ability of the latter to resist heat. Another important aspect which is taken into consideration is Gram staining reaction.

Besides morphological characters, other parameters taken into consideration for the identification of bacteria are the following cultural characteristics :
  (i)   Growth behaviour in nutrient broth medium
  (ii)   Growth characters on nutrient agar plates
  (iii) Type of growth on agar slants and
  (iv)   Type of gelatin liquefaction
Physiological characters of bacteria is another important parameter for identification of bacteria. Some of such important characters are enumerated below :
  (i)   Carbon utilization
  (ii)   Nitrogen utiliztion
  (iii) Relation to free oxygen
  (iv)   Type of by products released in the nutrient medium
  (v)   Reaction in milk and
  (vi)   Abilities of denitrification

Bacteria

Bacteria are the smallest and most abundant organisms in soil. They are free living and inspite of their size and simple morphology, they are infact the most interesting and marvellous component of the soil microbial population. They have been intensively studied and the group has added significance because of their involvement in the nitrogen and carbon cycles, and also in other cyclical transformations in soil. Though very tiny in size, bacteria have wide environmental tolerance and have a well established biotic relationships with other microorganisms or with higher plants.

Numerically the bacterial cells in soil are always great, and inspite of their small size, they account for a little less than half of the total microbiological tissue of the soil. Generally in well aerated soil, bacteria and fungi dominate the microbiological spectrum of the soil. However, in soil with little or no O_2, the major biological and chemical activities in soil are primarily the responsibility of bacteria. The group has certain advantages because of their small size, rapid growth and their ability to decompose a wide variety of natural substances rapidly.

Winogradsky (1925) classified the bacteria in two broad divisions i.e. the autochthonous or indigenous species, and the zymogenous or fermentation producing microorganisms. The categorization has an ecological rather than a scientific basis. The autochthonous population comprises numerous indigenous bacteria whose abundance is relatively fixed or stable in the soil. They derive their nutrients from the organic fraction of the soil, and they do not require an external nutrient supply or energy source for their survival and growth. Zymogenous organisms are relatively much less but they are the most active group in the chemical transformation. There is rapid build up in the zymogenous bacterial population when organic nutrients are added to the soil and as such, they respond quickly to the soil amendments and their population remains high as long as the externally supplied nutrients are available in the soil. Lateron, with the exhaustion of the organic substances the population drops down rapidly.

A more scientific and systematic classification of bacteria has been proposed in Bergey's manual of determinative bacteriology. This classification is basically taxonomical. There are other systems of bacterial classification also available and in most of such systems, a variety of nutritional and metabolic parameters have been used. Some of them are - nature of energy source, the carbohydrates used for growth, the capacity to utilize N_2 as nitrogen source etc. O_2 requirement of bacteria is another trait which has been used for the classification. Based on O_2 requirement bacteria may be aerobes, in which case O_2 is a necessity, anaerobes for whom O_2 is not at all required for growth, and for faculative anaerobes the development is possible either in absence or in the present of O_2.

Characterization of bacteria is also done on the basis of their cell structure. On this basis bacteria are categorized into three types :

(a) Bacilli - rod shaped bacteria, they are the most numerous in soil.

(b) Cocci - spherical shaped cell, and

(c) Spirilla - Spiral shaped cells, such type of bateria are not commonly formed in soil.

Under unfavourable conditions some of the bacilli form endospores which function as normal bacterial cell. Endospores are resistant, thick walled structures, and they help in better survival of the bacteria under prolonged dessication and high temperatures. Endospores persist in dormant state and help in the survival of bacteria under such conditions when normal bacterial cells find themselves unable to survive. On the return of favourable conditions, spores germinate and give rise to new organisms.

Bacterial Population in Soil

The population of bacteria in different types of soil is very variable both in terms of space and time. Normally in one gram of soil the number varies from 1 million or less to several billions. However, the different method used for the enumeration of bacterial number are far from satisfactory and therefore the bacterial population can not be measured with precision. The dilution plate method widely used for bacterial assessment, does not give the real picture on account of many lacunae. No single medium is suitable for the growth of all the different types of soil species. Besides this, the primary assumption in this method is that each viable bacterium in the soil suspension used as an inoculum develops into a viable colony. This assumption, however, is very unrealistic. Even if the figures calculated may be taken as approximately correct, they fall much below those obtained by using certain modern techniques. Strugger (1948) with the help of fluorescent microscopy, suggested the viable cells/g to be between 2-9 billions. Similarly, Taylor (1936) used the dilution ratio method for direct counting, and he observed a soil bacterial population of approximately 3 billion/g. By and large, it has been noted that the direct methods account for 10-50 times more bacterial cells in soil than that of agar plate counts. A survey of literature on the bacterial cells by direct observations reveals that usually the population is of the order of 1-3 billion/g. This is, however, an average figure and holds good for ordinary soil. The number may rise drastically in highly fertile soil, and particularly when external supply of the organic substances is made to soil. This increase is generally of a short duration and subsequently the population tends to attain stability.

Bacterial biomass

Pelczar and Reid (1958) suggested that one trillion bacterial cells, weigh 1 g and a direct count estimate of 2 billion bacterial cells /g of soil, accounts for approximately 0.2% of the soil weight. Based on these assumptions it is calculated that 4000 lb live weight of bacteria /acre are normally available. This calculation is, however, not ideal and variations have been observed in different types of soil. Many workers prefer a general range of values when the bacterial tissue per acre is conceived. Alexander (1961) estimated a range varying from 300-3000 lb/acre furrow slice, Russel (1950) from 1500-3500 lb, and Krasilnikov (1944) as much as 600-1200 lb for barren soil in uncropped soil, whereas the value for leguminous cultivated fields was noted to be in the range of 3600-6400 lb.

It has generally been observed that in most aerable soils, the quantity of living bacterial tissue per acre is somewhat less than that of the fungi. However, the same is much higher than those of algae, protozoa and nematodes combined.

Bacteria are more predominant in the water film of the soil. Theoretically speaking, in soil with 0.3 ml water /g, a bacterial population of 2 billion will occupy around 1.33% of the available space. Taking this fact into account in 0.3 ml there would presumably be about 150 billion cells. In case, there is no restriction of any type for the growth of bacteria and they are allowed to multiply, they will pack up the total porosity in soil.

In such a situation, the approximate figure for bacterial cells per g of soil will be 500 billion - which will occupy 0.4% of the available space. From the consideration that the available space occupied by bacteria is approximately 0.4 to 1.0% soil, theoretically this as such does not appear to be a suitable environment for bacteria. However, in reality the situation is different and it is convincingly proved by many studies that soil does provide an excellent environment for bacteria.

The only problem which is faced in the assessment of bacterial population is, the use of different techniques, which provide varying fluctuating results. Plate counts usually used for the sake of easy handling provide values ranging from several hundred thousand upto one hundred million bacteria per g of dry soil. This method, however, probably underestimates the true bacterial population density because the fact is that many bacteria fail to develop upon conventional media. On the other hand, direct microscopy estimates values to the tune of 10^8 to 10^9 bacteria /g of dry soil. Numerous techniques for direct observation of bacteria in soil have been suggested by different workers. Conn (1928) suggested a technique in which 0.015 per cent gelatin is used for suspending the soil aliquot, which is later on spread over a calibrated area of a microscopic slide and subsequently stained with rose bengal and observed under the microscope, preferably under oil immersion objective. Jones *et al.* (1948) proposed a modified technique which is more refined. In this technique, a weighed amount of soil is incorporated with melted agar and drops of agar infusion is added. The preparation is then placed on a calibrated haemocyto meter. The suspension is finally stained and examined under a microscope. Through this method it is possible to assess the bacterial quantity, provided, the amount of soil, the volume of the gelatin or agar and the area over which the smear is spread are known.

For qualitative studies the method developed by Rossi and Cholodny, which is more appropriately known as Rossi-Cholodny buried slide or contact slide method, is very useful. In this method, a microscopic slide is buried in the soil and after a desired time the slide is carefully removed and made free of extraneous materials like larger debris etc. The slide during burial period is impregnated with microbial film adhering to it. This is then washed carefully and stained before microscopic examination. This method has its own advantage because it allows microorganisms to develop in the physical posture typifying their normal position and associative relationships with their neighbours. This is an excellent tool for the ecological studies of microorganisms and has the added advantage that one can assess the associations and inter relationships between microorganisms and their counterpart soil particles, plant roots and other organisms.

It is interesting to note that there is no correlation between plate and direct microscopic methods. The values obtained through direct microscopic studies have also been questioned and it has been argued by some workers that higher population estimates result from accumulated dead cells (Topping, 1938). It, thus, appears that where as the viable count is somewhat low because of the inadequacy of the culture media, direct counts are exceedingly high because many non-viable individuals are included in the estimates. Thus the situation about real bacterial assessment in the soil still remains a controversial matter and it is left for future researchers to dwell upon the issue. It is possible that in future, if more sophisticated and sensitive techniques are developed, they may provide more realistic and satisfying results.

In addition to the above methods to assess the bacterial biomass, there are indirect methods also which are mostly based on the activities of bacteria. Number alone, many a time does not indicate the microbial activity. It is possible that many a time, particularly when the population estimate is made by direct microscopy, the bacterial propagules may comprise largely of dormant or resting bodies. In such a situation, it is advisable to assess the general activity of the soil microbial population and measurement of the respiratory activity is one such method. This is assessed in terms of CO_2 evolution. Different figures are available on such studies in the field condition. Alexander (1961) obtained 20 lb/acre per day CO_2 production in field soil. Vine *et al.* (1942) observed $3 \, 1/m^2$ per day where as Krasilnikov (1958) got a figure of 2 Kg/hectare per hour. Expressed on the basis of lb/acre per hour, these values become 0.83, 2.2 and 1.78 respectively - and thus the mean value is 1.6 lb CO_2/acre per hour. There always appears a discrepancy between the values obtained through cultural studies and those from the field. But with all the parameters taken into consideration, it is plausible that a major proportion of possibly 2 billion bacteria in a g of soil, are in resting or dormant condition and their respiratory activity is much lower than those of broth cultures. This assumption holds good mostly in the stationary phase of the life cycle of bacteria. From the above discussion it may be concluded that respiratory values and population considered together, give a clue that soil permits the survival of a large number of bacterial cells, many of them, however, being not very active metabolically.

Pattern of distribution in soil

There is no uniform pattern of the distribution of bacteria in soil. Even in the same horizon of the soil the pattern is haphazard. The distribution is, however, closely linked with the occurrence of organic matter. It has been discussed earlier that A horizon of soil is rich in organic matter and this layer accordingly harbours the maximum bacterial population. In horizons B and C, where the level of organic matter is low, the population of bacteria also is not much. Most of the studies confirm the above generalization of Starc (1942). On rare occasions, the organic matter concentration and the bacterial population do not correspond to each other. This generally happens when in A horizon the water content is low or there is too much acidity and then the bacterial number may be higher at lower horizons of the soil. In the soil where plants are growing, the root also affects the distribution of bacteria. Organic exudates and the sloughed off material of root surfaces provide an abundant source of energy material and this results in the luxuriant growth of microorganisms in the soil surrounding the roots. The root surface environment known as rhizosphere is a very suitable site for microbial growth and this aspect has been discussed separately.

Sometimes, islands or focii of microbial activity in soil have also been noted. Smith and Humfeld (1930) observed that in soil where green manure is plowed down, the bacterial population is very rich in the localized region where the green plant material is available in plenty, but a region very near to the plant material but free from the green tissue, exhibits very low population. Alan Burges (1967) also noted that in cotton fields, decaying cotton carpels in natural microclimates at the soil surface may support bacterial population much higher than in the soil away from the carpel deposition. The development of bacterial focii has also been noticed around the faecal droppings of insects and with dead mesofaunal and microbial tissues. Bacteria have also been seen to be associated with disintegrating strands of fungus mycelium. Alexander and Jackson (1954) noted that bacteria in soil are generally present in the film of colloidal material coating the mineral particles. It has been observed by Jones and Mollison (1948) that bacteria occur in small colonies in soil. They also suggested that 77% of the soil-bacteria occur in groups and the remaining 23% appear as single cells. However, Minderman (1956) noted that the majority of bacterial cells occur as single cells.

There is convincing evidence to show that bacterial cells do occur as single cells in soil and in such condition, they are transported from one place to another in soil through various transporting agencies like gravitational water and hyphal extension of fungi. It is suggested that in case of filamentous fungi particularly in fast growing ones, when they grow further in the soil they also carry the bacterial cells along with them on their body. Air borne dust and machines used for agricultural purposes also help in the transportation of bacterial cells. Certain mobile bacteria propel themselves in the water films present around the soil particles. With the help of the above transporting devices bacteria are carried to the different soils.

Bacteria predominantly found in soil

The most abundant bacterial forms present in the soil are of the coccoid rod

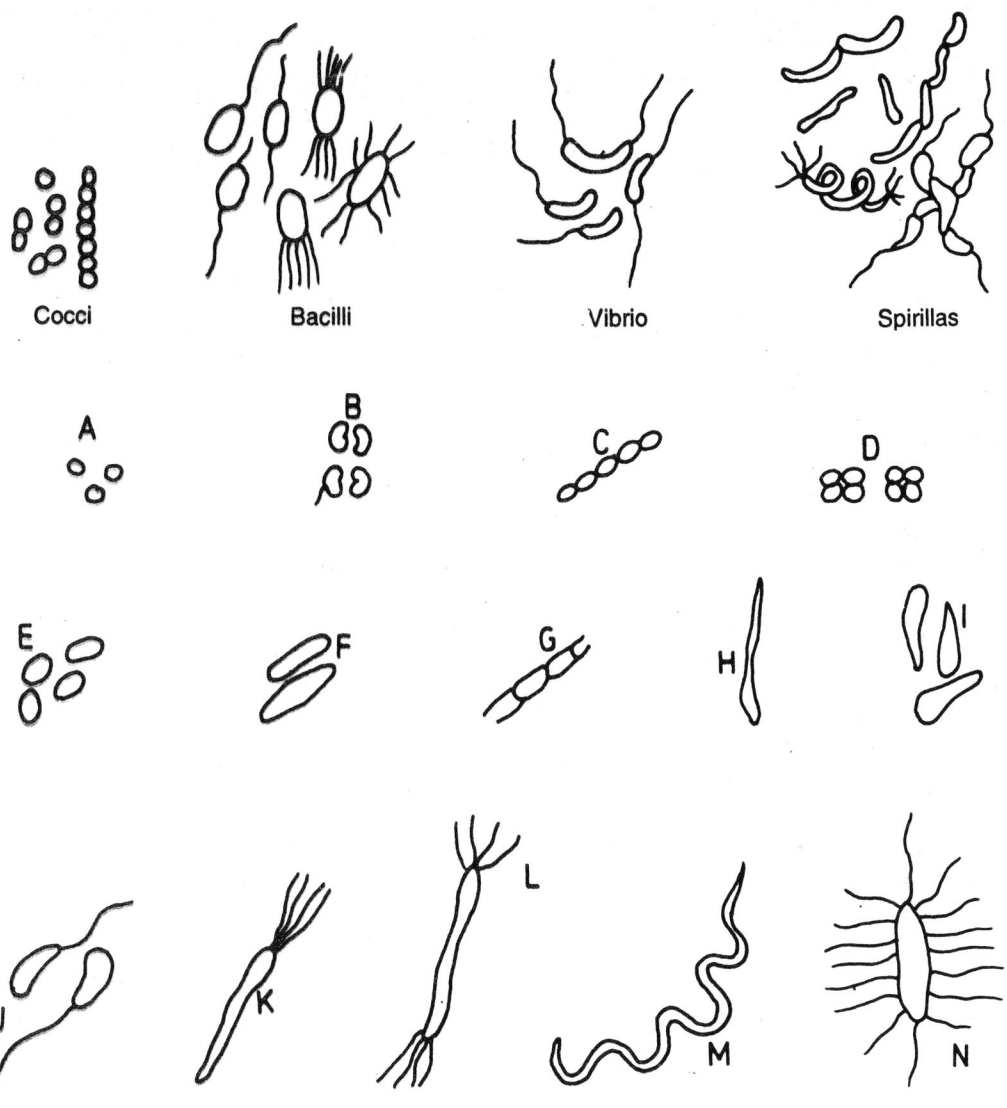

Cocci   Bacilli   Vibrio   Spirillas

Different Forms of Bacteria

A. *Staphylococcus*   H. Fusiform bacilli
B. *Diplococcus*   I. Clavate bacilli
C. *Streptococcus*   J. Monotrichous vibrio
D. *Micrococcus*   K. Lophotrichous
E. Small bacilli   L. Amphitrichous
F. Large bacilli   M. Spirillum
G. Chain bacilli   N. Peritrichous

Fig. 1.1(b) Different forms of bacteria

type. Cells of such type, in the young stage in cultures, are usually Gram negative rods. In rare cases they exhibit rudimentary branching also. Later on, however, cells of the older colonies assume cocci nature and they are Gram positive. Thus there is evidence of Pleomorphism. Majority of the soil bacteria of globiforme type are placed in the genus *Arthrobacter*. Breed *et al.* (1957) placed the cellulose decomposing species in the genus *Cellulomonas*.

Burges (1958) grouped the soil bacteria into the following six groups:

(1) Small cocci about 0.5 μ in diameter;
(2) Short rod about 0.5 μ in diameter, 1 to 3 μ long;
(3) Short curved rods, the Vibrio;
(4) Long rods;
(5) Rods sometimes showing branching;
(6) Thin flexible rods with very thin walls, usually under 0.5 μ in diameter.

As such *Arthrobacter* may include the types grouped in 1, 2, 4 and 5 categories and also sometimes of group 3 and 6.

Besides the above sporulating bacilli, the actinomycetes are the other group which are abundantly present in soil. With information available in the literature, it may be safely said that through cultural enumeration, sporulating bacilli constitute approximately 25% of the total number of bacteria in soil.

Mycobacteria constitute the group which is commonly found in soil. However, population density of this group is less than the corynebacteria or the bacilli.

From the survey of literature it may appear that coryneforms account for as much as 65%, followed by bacilli 25%, then Pseudomonadaceae, Nitro-bacteriaceae, Rhizobiaceae, Azotobacteriaceae, Achromobacteriaceae and Micro-coccacceae combined together contribute not more than 10% of the total bacterial flora. Genera wise *Agrobacterium*, *Azotobacter*, *Nitrosomonas*, *Nitrobacter*, *Rhizobium*, *Pseudomonas*, *Achromobacter* and a few others are quite dominant in soil.

Holding (1960) suggested the occurrence of *Pseudomonas*, a Gram negative genus in soil. The genus is present both in the root region i.e. rhizosphere and in plant free soil. The population density of the bacterium as suggested by Holding was a little more than 5% of the total count. Clark (1940) recorded that fluorescent bacteria or *Psudomonas* and *Xanthomonas* combined to form about 1-30% of the rhizosphere population of various plants.

Factors affecting bacteria

It is abundantly clear from the above statement that soil in general is quite rich in bacterial population, but it is also observed that many of the bacteria are either inactive or show a very low level of activity. This compels one to think what might be the reason(s) for their limited activities. It is not that easy to give any specific reply to this question, because, in soil, where a number of factors are acting and interacting simultaneously to pinpoint a single factor as limiting is not advisable. However, some of the factors which may be listed as limiting are available food supply and certain physical and biological factors in the environment. Moisture, aeration, reaction and temperature are the major physical factors.

Availability of suitable food in the soil is of paramount significance for bacterial growth. By and large, in soil the food supply is not regular and fluctuations in the quality and quantity of the food is a regular phenomenon. Generally in uncultivated soils, litter provides the basic food material and with certain exceptions, the addition of litter to the soil is for a short period of the year. In cultivated soil, the situation is a little different, where besides litter in the form of crop remains, extraneous organic and inorganic substances are also supplemented to the soil for better growth of the plants. These additional inputs, either in the form of litter or chemical substances, abruptly enhance the bacterial population in the soil for varying periods, depending upon the availability of surplus food. This situation continues for sometime and when the food supply is exhausted or reduced, the bacterial population drops down and a stability is again obtained. The nature of the added energy material influences both the immediacy and the duration of the rise in activity, as well as the specificity of the corresponding flora. Most bacteria are heterotrophic and they use organic compounds synthesized by autotrophic microorganisms and higher plants, both for their energy requirement and cell carbon. Autotrophic bacteria, on the other hand, use CO_2 as a source of cell carbon, and energy is secured through inorganic oxidation. Heterotrophic bacteria largely depend on the organic matter added to the soil by the activity of higher plants and in this context, the contribution of autotrophic bacteria is almost negligible. It is interesting to note that though a vast amount of organic matter is produced annually and a substantial part of the same is added to the soil through efficient activity of the heterotrophic microorganisms, the annual rate of decomposition balances out quite nicely with the annual rate of production of organic matter. For the maintenance of this balance it is not necessary that the soil microorganisms should work with full capacity throughout the year. Rather, as indicated earlier, this capacity is for a short duration and for soil bacteria, surprisingly the food material in soil is perennially inadequate. It is obvious that in the short varying period of their activity, bacteria along with the other microorganisms are able to utilize the added substrate and maintain the balance.

It is also possible that in a situation where the supply of energy yielding substrate is adequate, short supply of one or more of the essential mineral nutrients or necessary growth factors, may be limiting for bacterial activity. For the decomposition of carbonaceous substrate in the soil, amongst the mineral nutrients, nitrogen is the most important. In case crop residues contain 1.5% or more of nitrogen, additional nitrogen is not needed for bacterial activity. However, where the nitrogen level is low, bacteria involved in decomposition need extra nitrogen. Further, nitrogen requirement is more particularly during the early stage of decomposition. However, in normal aerable soil, the amount of nitrogen available is in the range of 20 to 100 lb, which is sufficient for bacterial activity. It is normally a very balanced situation and the quantity of residue annually added to the soil has a close relation with the nitrogen fertility of the soil. In most cases, the nitrogen content of the residues coupled with available soil nitrogen, is adequate to meet the nitrogen demand of the soil organisms involved in decomposition. As such, addition of nitrogen rarely accelerates residue decomposition in the field. However, when the situation is evaluated in terms of the nitrogen requirement of the crops and the nitrogen utilization by bacteria, the scenario may be different. It is quite possible that sometimes the soil organisms may demand all the available soil nitrogen in the presence of abundant energy material, and then the soil may be left with less nitrogen for the crop and hence nitrogen fertilization may become essential for the crop but not for bacteria.

Moisture requirements

Moisture is one of the most important factors governing bacterial activity. It does so in two ways : water is the major component of the protoplasm and hence it is essential for the vegetative growth and development of bacteria and its steady supply must be maintained. When moisture becomes excessive, the microbial population is decreased and accordingly the activity of the microbes is also minimised. Excessive water leading to the water logging situation doesnot favour the growth of microbes directly and also the effect is indirect through depletion in O_2 level and simultaneous increase in CO_2 concentration. The gaseous exchange in water logged soil is adversely affected. O_2 is not allowed to enter in the soil and CO_2 released through microbial activity and respiration of the under ground plant parts, gets accumulated in the soil, creating an environment in which aerobic forms do not thrive well and only the anaerobes grow and multiply.

The maximum bacterial density is found in regions of fairly high moisture content and the optimum level for aerobic bacteria is normally between 50 to 75% of the soil moisture holding capacity. Some bacteria are more active in drier conditions and not all bacterial transformations are uniformly curtailed during drying out of soil. It has been reported that ammonification can proceed under stringent drought than can nitrification. Robinson (1957) observed that ammonification occurs in soil with a moisture content one half the wilting percentage. Generally, however, there is a marked reduction in the population of bacteria in soil as it undergoes drying. Certain bacterial species are resistant to drying and they form the bulk of the population. Soil bacteria on an average are resistant to such extreme environments. This is the reason why, once the soil gets re-wet, the resistant bacteria present in it become active and they participate in varied phenomenon like nitrification, ammonification, non-symbiotic nitrogen fixation and sulphur oxidation. Calder (1957) observed that nitrate productivity of dry soil stored for 3 years was almost unimpaired. Similarly Sen and Sen (1957) noted that *Rhizobium japonicum* could survive 19 years during storage of the soil in an air-dry condition. Clark (1967) also reported the survival of *Azotobacter* in soil which was stored in air dry condition for 30 years. The findings, however, should not be taken as a generalization, because in many cases, when the soil has been subjected to excessive drying, this leads to the elimination of most of the bacterial species. *Rhizobia* may be cited as one specific example which fail to withstand soil drying. It is, therefore, always advisable to inoculate the seeds of legume with *rhizobia*, where the annual cropping of the legumes is practised.

Aeration

Bacteria are more predominant in properly aerated soil. Most bacterial species need a sufficient supply of oxygen. Generally soil contains more than 10% oxygen and in most-well-drained soils, the value may be above 20%. In wet periods or following heavy irrigation (Kemper and Ameniya, 1957), the oxygen content in soil is around 3% or even less. Movement of oxygen into the larger pores of the soil atmosphere is achieved by gaseous diffusion. The oxygen pressure difference necessary to cause adequate movement through air-filled pores needs equal only 1-4% oxygen. The situation is not so simple in soil where the bacteria are surrounded by water films and in such a situation, the rate of oxygen diffusion through water is only about one ten- thousandth as fast as the rate through air. Water thus becomes a barrier for O_2 movement and this leads to a limiting situation for bacterial respiration. Such a situation arises either by heavy rains or by irrigation practices. When soil is covered with water for a longer period, the amount of O_2 gets depleted, because the amount in any case is not inexhaustible and the O_2 initially entrapped in soil is used by the microorganisms present there. The disappearance of O_2 depends on the rate of oxygen used by soil organisms and plant roots and the amount of O_2 initially present. In many soils, however, the entrapped O_2 is sufficient to supply the biochemical oxygen demand from several days to a week.

Well drained soils are normally fully aerated throughout the entire profile, more so in case of droughty surfaces. In water sealed situations, either through sufficient precipitation or irrigation, the plow layer loses its O_2 more rapidly than does the deeper profile. This is because of the greater population of microbes and thus the greater biochemical O_2 demand in the top- soil. Under such conditions the subsoil is better aerated than the topsoil.

Many a times barriers that occur in and around individual soil pores and bacteria, pose a much more difficult problem because these barriers can restrict the movement of O_2 to bacterial cells. Sometimes it has also been observed that microsites of anaerobiosis occur even at the top layer of the soil and this provides enough opportunity for anaerobes to proliferate at the upper layer. With these observations in mind, it can be concluded that, along with aerobes which are more dominant at the upper layers of the soil, anaerobic bacilli also are distributed throughout the soil profile.

It may be safely concluded that, within a range of 50-80% of moisture holding capacity, soil is endowed with good soil aeration. In such wet condition, a good combination of moisture and aeration occurs, that favours the growth and activities of the heterotrophic bacteria in soil.

Reaction and Temperature

Bacteria normally prefer the alkaline side of neutrality. The common range for different bacteria is between pH 4 to pH 10. Some species show a wide tolerance to extreme reactions. Such variations are even exhibited by different species of the same genus. *Azotobacter chroococcum* flourishes both in acidic and alkaline soils, but rarely below pH 6. *A. indicus* on the other hand, had been isolated from very acidic soils having pH 3. Similarly *Thiobacillus thiooxidans* tolerates an acidity of pH 6. By and large, highly acid or alkaline conditions tend to restrict many common bacteria. It has been observed that, the greater the hydrogen ion concentration the smaller is the size of the bacterial population.

Temperature

Like other parameters, bacteria have wide aptitude in relation to temperature. Optimum temperature range for soil bacteria is between 25-35°C. However, a large number of bacteria may be isolated over a range of 10-40°C. In normal conditions, such extremes are seldom seen in soil. Usually high temperatures occur at or near the surface in dry and barren soils. In tropic soil, particularly in summer months, the bacterial population is affected by high temperatures especially at the upper surface. At lower levels of soil where the roots are plenty, even in drier situations, the soil temperature is usually either optimal or sub-optimal for bacteria.

As in other organisms, in bacteria too biological processes are governed by temperature. A close association between population, size and temperature has been noted by many workers (Jensen, 1934). On the basis of temperature tolerance for their growth and proliferation, bacteria may be grouped into 3 categories :

(a) **Mesophiles** — Those with an optima in the vicinity of 25 to 35°C and with a capacity to grow between 15-45°C. Most of the soil bacteria fall under this category.

(b) **Psychrophiles** — Certain species prefer low temperature say below 20°C. True psychrophiles are generally not present in soil. In severely cold months, when the temperature falls below 20°C the bacteria isolated from soil at such a time, are not really psychrophiles but cold tolerant mesophiles.

(c) **Thermophiles** — Bacteria growing between 45-65°C are obligate thermophiles. Such species usually do not grow below 40°C.

In thermal hot springs or during volcanic activity, temperatures above the optimum occur in soil or water. Sufficiently high temperatures have also been recorded in stored hay and grains.

Besides, the factor enumerated in the preceding pages, certain other soil characteristics also affect the soil bacteria. The application of fertilizers in cultivated fields is one of the important factors which govern the bacterial population. Application of mineral fertilizers serves a dual function. They provide needed minerals both to the plants and microorganisms. Bacteria have also been suppressed by the use of ammonium fertilizers. The suppression effect is therefore not directly due to the added nitrogen but rather indirectly as a result of the acidity generated through the microbial oxidation of ammonia to nitric acid.

Cultivation of the soil is also an important factor in soil bacteriology. Different agricultural operations like plowing and tillage have a marked influence on the soil bacteria. These agricultural operations improve the soil structure and porosity, facilitating free gaseous exchange in soil. The moisture status of soil also changes and the hitherto inaccessible organic nutrients are exposed for bacterial action. The bacterial population under such changed environments increases.

Seasonal climatic changes have a pronounced affect on soil bacteria. The effect is more so in tropical soils. During summer months when the soil temperature rises, the moisture regime of the soil gets lowered. These two variables result in the low soil bacteria density. With the onset of rain, the temperature drops and the moist soil becomes more conducive for the bacteria. An abrupt increase in bacterial activity is noticed and this continues during the rainy months. At times, a rather enhanced bacterial density is noted at the end of the rainy season, when the dead seasonal plants contribute significantly to the nutrient status of the soil. Once the added plant materials are exhausted the population of bacteria decreases and this tendency continues for the rest of the winter months, reaching the lowest peak in the hot summer days. The seasonal change in the number of bacteria is closely related to the various factors operating simultaneously during different seasons of the year.

Chemoautotrophic bacteria

As mentioned earlier, bacteria, depending upon their mode of nutrition, are largely heterotrophs. A few species, however, exhibit different nutritional behaviour and they may be loosely grouped as autotrophs. Unlike green plants which derive their energy from sunlight certain bacterial species obtain the energy needed for growth and biosynthetic reactions from the oxidation of inorganic materials and they are known as chemoautotrophs. Chemoautotrophy is limited to relatively few species of bacteria. The latter utilize the energy obtained by the transformation of inorganic material and have the capacity to make use of CO_2 to satisfy their entire carbon needs. Amongst chemoautotrophs too, two major groups may be identified. Obligate chemoautotrophs are limited exclusively to inorganic oxidations; others, like the facultative autotrophs, obtain energy from the oxidation of either inorganic materials or organic carbon. Physiologically, chemoautotrophic bacteria are very interesting and complex. They have within the confines of their cell walls, all the enzymes, vitamins, coenzymes, carbohydrates and other protoplasmic constituents typically found in heterotrophs. Coupled with all these characteristics, the synthetic powers of chemoautotrophs is truly great. This group, though numerically small, has great economic and agricultural importance. Certain very important genera like *Nitrosomonas nitrobacter, Ferrobacillus* and certain species of Thiobacillus, are representative of true chemoautotrophs. Most chemoautotrophs are strict aerobes. Other, capable of proliferating in the absence of O_2 require an O_2 rich substance, like nitrate for *Thiobacillus denitricans*, sulphate for *Desulfovibrio*, and CO_2 for *Methanobacillus*.

Some of the important energy yielding reactions related to such bacterial species are :

$$NH_4 + 1^{1/2}O_2 \longrightarrow NO^-_2 + 2H^+ + H_2O \text{ } \textit{Nitrosomonas}$$

$$NO_2 + 1/2 \text{ } O_2 \longrightarrow NO_3 - \textit{ Nitrobacter}$$

$$S + 1\text{ }1/2 \text{ } O_2 + H_2O \longrightarrow H_2SO_4 \text{ } \textit{Thiobacillus}$$

$$2H_2 + O_2 \longrightarrow 2H_2 \text{ } O \text{ } \textit{Hydrogenomonas}$$

$$4H_2 + SO_{4=} \longrightarrow S^= + 4H_2O \text{ } \textit{Desulfovibrio}$$

$$4H_2 + CO_2 \longrightarrow CH_4 + 2H_2O \text{ } \textit{Methanobacillus}$$

Chemoautotrophs may be subdivided as follows. This division is on the basis of the element whose oxidation provides the energy for growth and cell synthesis.

(i)   Nitrogen compounds oxidised
      A. Ammonimum oxidized to nitrite — *Nitrosomonas*
      B. Nitrite oxidized to nitrate — *Nitrobacter*
(ii) Inorganic sulphur compounds converted to sulphate — *Thiobacillus*
(iii) Ferrous ion converted to the ferric state — *Ferrobacillus, Gallionella*
(iv) H_2 oxidized — *Hydrogenomonas, Methanobacillus, Desulfovibrio*
(v)   CO oxidized to CO_2 — *Carboxydomonas*

Bacterial species belonging to chemoautotrophs are very important in nature. This is because of their ability to catalyze energy- yielding reactions and several of these processes are important in agriculture. The formation of nitrate and sulphate is useful for plants because it provides inorganic nutrients in assimilable form. The reduction of sulphate to sulfide, besides having an influence on plant growth, is also associated with the corrosion of iron and steel pipes in the soil.

Biological activity of soil bacteria

With the foregoing discussion it is clear that bacteria are essential for soil fertility. The major contribution of bacteria is to decompose organic nitrogenous compounds in plant and animal residues, with the ultimate liberation of ammonium. The latter can be oxidized to nitrate only by nitrifying bacteria. Such bacteria obtain their energy by the exothermic oxidation of ammonia to nitrite and of nitrite to nitrate. Chemoautotrophic bacteria are mainly responsible for such transformations, and they exhibit substrate specificity of a very high degree. No bacterial species is capable of carrying out both these oxidations in the conversion of ammonia to nitrate. Initially, Winogradsky in 1890, isolated representatives of these two groups and they were designated as *Nitrosomonas* and *Nitrobacter*. The situation is different now and five species in four genera are know to be ammonia-oxidizers and three species in 3 genera are nitrite oxidizers (Stanier *et al.*, 1977). All these species are Gram-negative.

Another important contribution of bacteria is in the nitrogen cycle. Certain bacterial species are unique in fixing atmospheric (molecular) nitrogen in the form of ammonium nitrogen. Molecular nitrogen is fixed only by pro-karyotic cells. Nitrogen can be fixed by bacteria either symbiotically, in which case they grow in the roots of the host plants and form nodules or by free living bacteria. In *Rhizobium*, a Gram negative genus, five species have been identified which fix nitrogen. Largely the species are confined to plants of the family leguminosae. Recently it has been reported that besides legumes, root nodules are formed in other plants also. Alder (*Alnus glutinosa*) is prominently reported to be with root nodules. The nodules are not bacterial, but the microorganisms identified are *actinomycetes*, which are grouped as the Gram- positive bacteria. This association has attracted the attention of many workers because of the nitrogen fixing ability of *actinomycetes* like *Rhizobium*.

The nitrogen fixed by free living bacterial species is relatively less. Two genera *Azotobacter* (aerobic) and *Clostridium* (anaerobic) are the best known nitroger fixers. *Azotobacter* is very common in the rhizosphere region of plants and they maintain themselves on the root exudates. The plants rich in *Azotobactei* population have been observed to grow better. Besides nitrogen fixation *Azotobacter* is also useful to the host plants through the production of gibberelins and possibly other growth hormones, which results in an increased crop yield due to bacterization (Brown 1974).

FUNGI

In contrast to soil bacteria, the studies regarding soil fungi are of relatively recent interest. Although knowledge on fungal fructification occurring on soil is ancient, the credit of isolating fungi from soil goes to Adametz (1886). A little later in 1902, Oudemans and Koning made a detailed study of soil fungi. They identified forty five species from the organic matter extracted from soil. Hagem (1907, 1910), Lender (1908), Dale (1912, 14) Bechwith (1911) and Jenson (1912) can be cited as being the pioneer workers in the field of soil fungi. Inspite of fragmentary information being available regarding the presence of fungi in soil there was still a lot of confusion and controversy as to whether fungi really live in soil. The issue remained a controversy for a long time. Waksman (1916 a) raised the question of whether soil is the home of an indigenous mycoflora or merely a resting place for fungal spores floating in the atmosphere. Later on, Waksman himself along with many other workers, argued that many fungi grow and reproduce in soil. However, soil is decidedly a sink for a wide variety of microorganisms from different habitats. For a number of fungi, soil appears to be little more than a resting space, for many others on the other hand soil provides a good milleau and they complete their life cycle there.

In well-aerated cultivated soils, fungi account for the largest part of the total microbial protoplasm. This dominance in mass is the result of the large diameter and the extensive network of fungus filament. In soil, the range of fungi is very wide from chytrids to agarics, from saprophytes to root rot, from parasites of amoebae to parasites of man. On account of the varied nature and activities, the fungi as a group have been the most interesting and widely studied amongst the various soil microorganisms. Cooke (1958) opined that soil has been studied more extensively than any other natural habitat of fungi. This is mainly because of the various roles the fungi play in soil. They are known to be plant pathogens, they are active participants in the decomposition of plant and animal residues and they are also involved in mycorrhizal association.

Apart from a few higher fungi like mushrooms etc., the presence of fungi in soil cannot be visualized with the naked eyes. Various methods are employed for the assessment of the fungal population in the soil. Based on studies, it has been demonstrated that fertile land may harbour as much as 10 to 100 m of active fungus filaments per gram of soil. Assuming the mold filament has an average diameter of 5 μ and a specific gravity of 1.2, Alexander (1961) suggested that the live weight of fungi is approximately 500 lb. per acre. This calculation is, however, an approximate and average and the live weight may considerably differ from one soil to other depending upon many factors operating in the soil.

Recently the ecology of soil fungi has attracted the attention of many soil microbiologists on account of their active role in the root region. The rhiosphere and mycorrhizal regions are the best studied environments in relation to the soil fungi. These aspects have been described in detail in specific chapters.

Much of the earlier work on soil fungal flora has been essentially floristic. The emphasis has, however, changed now and the ecology of soil fungi has now been the major field for detailed investigations. Also the habitats of individual species and the part fungi play in the biochemical processes that take place in soil are the other interesting and recent areas which have attracted the attention of soil mycologists. On the basis of recent studies, it is now abundantly clear that fungi in soil are mainly in the form of resting structures, in a mosaic of micro-habitats often making little mycelial growth. The situation changes rapidly when some event brings fresh nutrients to the resting cells and then there is intensive activity to exploit the freshly added nutrients. Some of the events responsible are root growth, litter accumulation and the activity of the soil fauna.

Inspite of intensive research being done in the various aspect of soil fungi during the last half a century, the knowledge regarding the spectrum of fungi present in the soil, is far from satisfactory. What we know, particularly with respect to the type of fungi present in the soil, is probably only 50%. There are still, presumably, many more fungi in the soil awaiting discovery. This lacuna is mainly due to the shortcomings in the methodologies adopted for the isolation and enumeration of the fungi from the soil. Like bacteria, in case of other soil microorganisms including fungi, we still have difficulty in identifying and perfecting the techniques which may be suitable for providing full information to our satisfaction. Numerous methods which have been suggested for such studies have their own limitations and so is the case for the information collected through them.

In general, there have been two different approaches for the study of soil fungi. The first is by microscopic examination either of soil or of substrates or materials such as glass or nylon after they have been placed in soil; the second is by the isolation of organisms either directly or by cultural techniques. Each approach has both advantages and disadvantages. In general, compared with some other habitats, soil is more difficult to study. This is mainly because of the multitude of organisms that occur in soil, of the complexities of fungal life cycles, coupled with the difficulties inherent in investigating soil because of its opacity, its hetergenous nature and its complex structure.

Isolation Techniques

Several techniques have been suggested for the isolation and study of the soil fungi. Some of the important techniques are given below :

A. Direct observation

(1) **Soil Sections**

An undisturbed soil sample is soaked in a hot, 2% aqueous solution of agar, cooled, hardened in alcohol and then sectioned as thin as possible. Normally 750 μ thick sections are reasonably good for sandy soils, while for organic soil the best are 100 μ thick. This method widely used for microscopic studies was suggested by Haarlov and Weis Fogh (1953, 55). Subsequently some modifications were suggested by others. However, for routine observations the one suggested above is the best.

(2) **Slide or burial method**

This method involves the processing of a clean microscopic slide against a freshly cut soil surface. This provides an opportunity to the microbial colonies to adhere to the slide. The slide is allowed to remain in the soil for some time and then removed and stained. Care is taken not to disturb the soil so that the slide can depict the microorganisms as they actually occur in the soil at the time of burial of the slide. Rossi (1928) was the first microbiologist to suggest this technique which, however, was perfected and made popular by Cholodny (1930). This subsequently become known as the Rossi-Cholodny or contact slide method.

(3) **Soil staining**

Conn (1918) was the first to stain soil suspensions. In this method an infusion of soil (1:9) in dilute gelatin is prepared and 0.1 ml of the infusion is spread across a slide. which is then stained with rose-bengal or erythrosin. Later-on Jones and Mollison (1948) suggested some modification. They suspended soil in melted and cooled 1.5% agar. A drop of the agar suspension was placed on a haemocytometer and a cover slip quickly placed over it. A film was obtained which was floated off on sterile water. It was then placed on a microscopic slide, allowed to dry and stained with phenolic aniline blue. This preparation, if needed, can be made a permanent mount. The advantage of this method is that the total length per gram of soil of hyphae may be measured.

The above three methods described under direct observations, provide an insight to the fungal propagules as they exist in the soil. Many a time it is not so easy to identity the fungi obtained because they might be in the form of hyphae, spores, sclerotia or other unidentifiable structures.

Isolation methods

The methods described below are conventionally used by soil mycologists for routine work. Soil from a desired place is collected and inoculated on nutrient plates and the fungi appearing on the plates are identified. The advantage of these methods is that the fungi so obtained can be readily identified.

(1)   **Soil dilution plate method**

Waksman (1927) suggested this method. In this method, a known amount of soil is shaken in sterile water. From the soil suspension, a progressive series of dilutions are made. 1ml of the final suspension is then dispersed with melted but cooled agar in a Petri dish. Replicates may be made depending upon the need. Many a time, the growth of fungal colonies is checked due to the rapid appearance of bacterial colonies. To avoid this problem Dawson (1944) proposed the use of rose-bengal at a conc. of 1 : 15,000 as a bacteriostatic agent. The use of antibiotics 30 g/ml streptomycin or 2 g/ml aureomycin in combination with rose-bengal has also been recommended (Martin. 1950).

By soil dilution plate method, the number of colonies /g of oven- dry soil, can be calculated.

(2)   **Soil plate method**

Warcup (1950, 60) suggested the use of a small quantity of soil directly in place of the soil suspension used in the above method. The soil is dispersed through a thin layer of agar medium in the isolation plate.

The advantage of this method is the ease of preparation and this technique also provides an opportunity for growth of the hyphae associated with plant residue which are other wise discarded in dilution plate method.

Immersion tube technique

A specially designed immersion tube, made of a glass or plastic cylinder, with a small opening in its side is used in this technique. The tube is filled with a suitable solid medium and sterilized. A small hole is bored in the soil and the apparatus is inserted tightly in it. The immersion tube is left in the soil hole for about a week and brought to the laboratory for the isolation of the hyphae developing through the agar (Chester, 1948).

The above stated techniques have some edge over the one on direct observation, but there are a lot of shortcomings too. For the purpose of enumeration, conventional plate counts have been most widely used and this permits the greatest degree of quantification, although the results are far from unequivocal . Estimates made on the basis of plate counts are open to serious criticism. We know little about the source from which the colony appears in the plates, whether they originate from spores or hyphae, in the plates and also the active or dormant nature of the viable unit in the soil sample, is unknown. Many a time readily sporulating genera like *Aspergillus, Penicillium, Mucor* and *Rhizopus* etc. appear in large numbers because of the numerous spores, each of which may give rise to a colony.

Also, the culture medium is nutritionally inadequate for the growth of all the fungal species present in a given sample of soil. Some times many species are not represented on the dilution plate because their propagules are inhibited at an early stage by the competition from fast growing fungi. Thus many slow growing species are unable to form visible colonies. It is a common sight to observe members of higher Basidiomycetes, like mushrooms, in soil during the rainy season but they are rarely found on soil dilution plates.

Looking into all these disadvantages, Warcup (1955 a, 1957) made a notable contribution to the ecology of soil fungi. He made special efforts to isolate soil fungi from the propagules associated with the debris of plant material which normally settle down at the bottom of the container wherein the soil suspension are prepared for dilution plate method. Warcup was able to identify the point of origin for about 95% of the colonies. Interestingly about 75% were able to arise from spores, 20% from within humus particles and 5% from visible fragments of hyphae.

In contrast to the dilution plate method, Warcup's soil plate method provides better information by incorporating the whole soil with agar. The method permits isolation of fungi that are otherwise rejected with the soil residues in the dilution plate method. This method elucidated by Warcup has, therefore, been widely accepted by soil mycologists.

To sum up, it may be said that not a single technique described above is perfect and free from criticism on all counts. Each of them have their own merits and demerits and the result obtained through them are not completely satisfactory. It is always advisable to follow different techniques simultaneously for a better overall picture of the fungi present in the soil.

Type of fungal structure in soil

Fungi are present in different forms in soil, ranging from unicellular forms to a much branched filamentous structure. The different species found in soil are drawn from different families and orders, they vary in size and life cycles. Some fungi live in soil purely in the vegetative form, some of them don't have reproductive structures at all and they are simply grouped as sterile mycelia. On the contrary, many form complex sporophores, particularly members of ascomycetes and basidiomycetes. It will not be out of place to describe the various forms which are exhibited by different soil fungi.

(a) **Hyphae**

On examination of soil by different techniques various types of hyphae are frequently noticed. Some of them are quite active and grow; their cells are full of cytoplasm and they possess growing tips which stain deeply. On the other hand, many hyphal structures lack content, do not stain appreciably and they apparently look collapsed and dead. Sometimes, it is not easy to identify if the hyphae in soil are living or dead.

Hyphae occurring in soil are of various types : Non-septate phycomycetous hyphae, fine or wide, hyaline or dark coloured, septate hyphae, hyaline or coloured hyphae with clamp connections. Amongst the various types of hyphae found in soil, it is difficult to assess how many of them are viable. There are various conflicting reports available on this matter. Warcup (1957) working on wheat-field soil, observed that on an average 23% of the hyphae are viable. It has also been noted that even for the same soil, the percentage of viable hyphae may vary at different times particularly in cultivated soil. Different stages of cropping may account for the diverse percentages in viability of hyphae.

The functional life of an individual hypha in soil is also a matter of controversy. Generally it is believed that hyphae are short-lived and sooner or later they are attacked by other organisms (Waksman 1927, Russel 1961).

Warcup (1957, 1960) observed that hyaline hyphae are generally short-lived, where as dark brown or black hyphae may be relatively long-lived. Amongst the long-lived hyphae many are "resting hyphae". Hyphae of vesicular - arbuscular endophytes and of some basidiomycetes too are long-lived ones.

(b) **Chlamydospores**

Chlamydospores are typically thick walled, often dark coloured resting spores. In most of the cases they are formed in vegetative hyphae, but sometimes they may also be a product of sporangiophores or conidia. Unfertilized oogonia may also result into chlamydospores. Various reasons may be assigned for the formation of such resting bodies. Non-availability of sufficient food material, high temperature, low C-N ratio and high concentration of sugars are some of the main factors for the formation of chlamydospores. *Mucor, Saprolegnia, Fusarium* and *Trichoderma* are some of the important soil fungi which form chlamydospores.

(c) **Rhizomorphs**

When the conditions in soil are not favourable, individual hyphae combine together and form mycelial strands. This is also common when hyphae are growing on the surface of soil. An examination of the litter commonly shows mycelial strands which are usually formed by some members of *Ascomycetes*, *Basidiomycetes* or *Hyphomycetes*.

(d) Sclerotia

Sclerotia are formed by interwoven hyphae which become globose or tightly compacted. They are generally thick-walled structures and are brown or black in colour. The size and shape of sclerotia vary with respect to the fungi. Sclerotia are rich in food reserves, particularly in oils and glycogen. Surrounded by a thick outer wall and enough food material inside, the structure has the capacity to withstand adverse soil conditions. A considerable number of fungal species occurring in soil are capable of forming sclerotia and with this adaptation they can survive under unfavourable conditions for a number of years. *Rhizoctonia*, *Phymatotrichum ominivoreum* and *Sclerotium cepivorum* are some of the common soil fungi which form sclerotia.

(e) Vesicles

Gerdemann and Nicolson (1963) reported that certain phycomycetous species form specialized types of vesicles which are widely spread in soil. These species are associated with endomycorrhizae. The population of vesicles in some soil is quite low and the functional importance of vesicles is dealt with in detail on mycorrhiza. It is interesting, however, to note that shape, size and colour of vesicles of different phycomycetes endophytes vary.

(f) Spores

Fungi as a group are very remarkable in the sense that they produce different types of spores during their life. Vegetative, asexual and sexual spores are frequently met with in most of the fungal species. It is through spores that dispersal of fungi mostly occurs in soil. In soil all sorts of spores produced by different fungi, varying in size, shape, structure and colour are encountered. It is difficult to estimate the longevity of the spores in soil. Generally, unicellular, hyaline, thin walled spores are short-lived where as multi-celled, thick walled, coloured spores are the long-lived ones. Contradictory reports are available in the literature about the life of different types of spores in soil.

In many cases it has been observed that there exist certain inhibitory factors in soil which inhibit the germination of fungal spores in soil. Dobbs and Hinson (1953) and many others later on (Hessayon, 1953; Jackson 1958) reported a widespread fungistatic factor in soil. Later studies suggested that this inhibition may be on account of nutrient deprivation, ethylene production or some factors operating in soil.

In addition to the spores widely present in soil, more complex fruiting bodies of ascomycetes and basidiomycetes are also commonly encountered in soil. Such fructifications have been observed in soil pores (Kubiena 1938, Warcup 1957), on roots and other structures in soil (Tribe 1957, Waid 1960).

Type of fungi in soil

Fungi belonging to all the major classes are widely encountered in soil. Members of the fungi imperfecti are numerically more frequent than the other groups. Depending on soil structure and the various physico-chemical characters, the types of fungi vary in different soils. No matter how close two locations in soil might be, there is never any close similarity in the composition of the fungi at the sampling sites. However, many a time when the isolation is made through conventional dilution plate method much similarity is observed in the type of fungi isolated from different soils. Burges (1958) opined that this apparent uniformity of the soil flora may be largely an artifact, based partly on uniformity in the method of isolation and partly on the fact that all the genera quoted have fairly characteristic sporing structures that allow easy identification at least of the genus, whereas isolates which are difficult to identify are seldom included in lists of soil fungi. Gilman (1957) gave a comprehensive list of fungi isolated from soil. He has listed about 690 fungi belonging to 170 genera. Amongst them 10 genera i.e. *Penicillium, Aspergillus, Mucor, Achlya, Pythium, Saprolegnia, Mortierella, Fusarium, Chaetomium* and *Monosporium* account for more than half of the species isolated. Gilman has listed 80 ascomycetes, 2 mycelia sterilia and no basidiomycetes at all. The following list summarises the common fungi isolated from soil :

I. **Phycomycetes**

    A.Mucorales.   *Absidia, Cunninghamella, Mortierella, Mucor, Rhizopus, Zygorrhynchus.*

    B. Pernosporales   *Pythium*

II. **Ascomycetes**   *Chaetomium*

III. **Fungi imperfecti**

    A. Moniliaceae   *Aspergillus, Penicillium, Acrostalagmus Botrytis, Cephalosporium, Gliocladium, Monilia, Scopulariopsis, Spicaria, Trichoderma, Trichothecium, Verticillium.*

    B. Dematiaceae   *Alternaria, Cladosporium, Pullularia*

    C. Tuberculariaceae   *Fusarium, Cylindrocarpon, Rhizoctonia*

IV. **Mycelia sterilia**

As indicated above, members of fungi imperfecti are more common in soil than representatives of other groups. Aspergilli and Penicillia are the maximum in soil as has been assessed by the dilution plate method. It has been reported that species of Aspergilli are abundant in warmer places whereas Penicillia are frequent in cooler places. As such, the former genus is particularly abundant in soils of the temperate regions, the latter in the tropics. Similarly, the relative incidence of *Rhizopus* and *Fusarium* seems to be greater in regions in close proximity with the equator (Mishustin, 1956). It has also have reported that Aspergilli are favoured in the dry praries

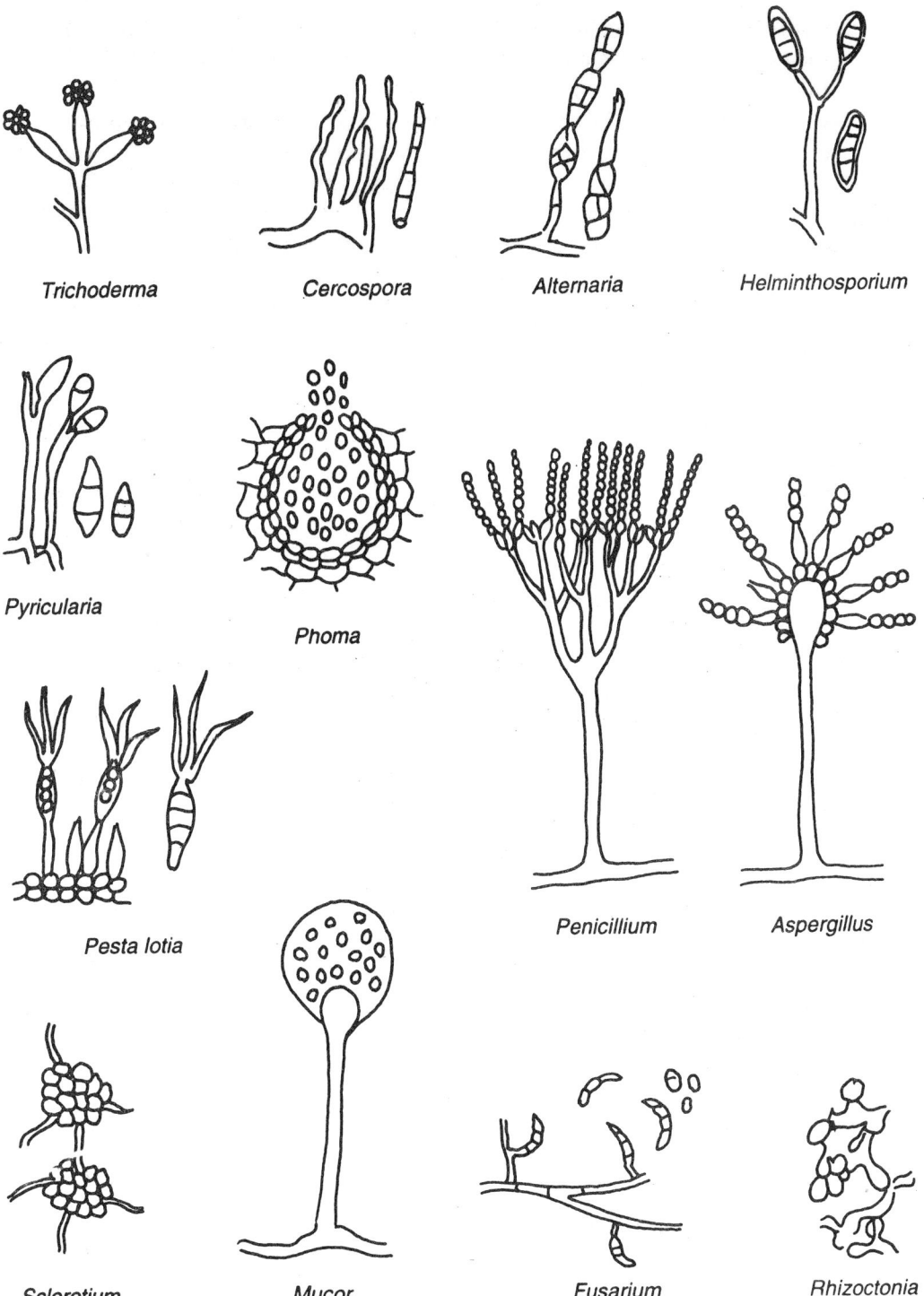

Trichoderma *Cercospora* *Alternaria* *Helminthosporium*

Pyricularia *Phoma*

Pesta lotia *Penicillium* *Aspergillus*

Sclerotium *Mucor* *Fusarium* *Rhizoctonia*

Fig. 1.1.(c) Common Soil Fungi

while representatives of the family Mucorales are of great frequency in wet locations. In woodland Mucor and Trichoderma are generally more numerous.

In addition to the above, selective methods have yielded other groups of fungi also. Chytrids (Willoughby, 1961), water- mould (Harvey, 1925), fungi which attack nematodes, protozoa and amoebae (Drechsler, 1941), plant pathogens (Garrett, 1956), animal and human pathogens (Ajello, 1956; de Vries, 1962) and mycorrhizal fungi (Harley, 1959) are some other important groups of fungi which have been reported by using selective methods.

Inspite of all this information being available regarding the isolation of several fungi from soil, it is not clearly known whether how many of the fungi complete their full life cycle in soil.

Fungi in soil are normally confined to litter, roots, plant residues, seeds, animal substrates and fungal structures themselves.

Myxomycetes or slime moulds are another group of fungi which are common in soil. Another group which is also wide spread in soil is yeast. Yeasts exist primarily as unicellular organisms and they reproduce by budding or fission. Two broad categories have been identified in yeast. The sporogenous group that produce ascospores and those that do not form ascospores. The ascospore forming yeasts are placed in the class ascomycetes and include genera like *Saccharomyces*, *Pichia* and *Hansenula*. The non- ascospore forming yeasts include *Candida*, *Rhodotorula* and *Cryptococcus*. The most common yeast genera enumerated from soil are *Candida*, *Cryptococcus*, *Debaromyces*, *Hansenula*, *Lipomyces*, *Pichia*, *Pullularia*, *Rhodotorula*, *Saccharomyces*, *Schizoblastosporion*, Torula, *Torulaspora*, *Torulopsis*, *Trichosporon* and *Zygosaccharomyces*.

Role of fungi in soil

Fungi in soil are associated with various types of activities. Some important ones are listed below:

(a) **Utilization of carbonaceous materials:** In this category five major groups have been identified: Saprophytic sugar fungi, lignin decomposers, coprophilic fungi, predaceous types and root inhabiting fungi.
   Members of phycomycetes are generally associated with sugar decomposition and they rarely decompose lignin and cellulose. Basidiomycetes as a group are lignin decomposers.

(b) **Utilization of proteinaceous substances :** Many fungi are active in the formation of ammonium and simple nitrogen compounds in soil. Various fungi participate in this process.

(c) **Plant pathogenicity**
   Certain soil fungi are common plant pathogens particularly root pathogens. Some saprophytic species act as pathogens and at an opportune time invade living tissues.

(d) **Mycorrhizal formation**

A number of fungi are associated with the root of the plants and they develop a specialized symbiotic association with the host plant, known as mycorrhiza. Two well known mycorrhizal association i.e. ectomycorrhiza and endomycorrhiza are the result of two groups of fungi. In the former case members of basidiomycetes and a few ascomycetes are involved, whereas in the latter, the major contribution is by certain members of phycomycetes. This aspect has been dealt with separately in chapter 2.

Factors which affect soil fungi.

Survival, adaptation and establishment of fungi in a particular soil is determined by a number of environmental factors. Organic matter status, hydrogen ion concentration, temperature, profile, moisture regime, aeration, season of the year and composition between vegetation are the important external factors determining the distribution of fungi in soil.

Organic matter : Distribution of fungi is largely determined by the availability of oxidizable carbonaceous substrates. As indicated earlier, the upper horizon of soil is rich in organic matter and correspondingly this layer is also rich in filamentous fungi. In general, the organic matter status of the soil decreases with depth and this has a direct bearing on the fungal population, therefore the soil at lower depths is poor in fungi. In natural soil, the organic status is generally stable and so is that of the soil fungi, where as in cultivated soil the situation is different. In cultivated soil the nutrient status is altered or rather improved by incorporation of crop residues, green manures or other energy rich carbonaceous materials and these amendments increase the size of fungal populations. Similarly in grasslands and in soil with annual plants, an increase in the fungal population is observed when the cover vegetation dies and litter is added to the soil. In tropical situations an abrupt increase in soil fungi in grasslands and cultivated fields is noted by the end of the rainy season when, besides sufficient organic nutrients, other soil conditions like temperature and moisture are quite conducive (Mishra 1960; Dwivedi 1966,68).

Hydrogen ion concentration

The acidic side of pH is generally more favourable to soil fungi. There are, however, many exceptions. Moulds, for example, develop over a wide range of pH. It is not uncommon to isolate species of *Aspergillus* and *Penicillin n* from wide pH ranges. Bacteria and actinomycetes are uncommon in acid habitats and there is a lack of microbial competition for food etc. Under such conditions, fungi naturally find themselves in an advantageous position to grow and multiply luxuriantly. Biochemical transformations in acid habitats are therefore largely as a result of fungi. In soil amended with inorganic fertilizers, particularly those rich in ammonium salts, the population of filamentous fungi increases. Such an increase is generally due to the acidification of soil on account of the microbial oxidation of nitrogen, which leads to the formation of nitric acid. Many a time, repeated annual addition of ammonium fertilizers, though favourable to the fungal population, results in decreased bacterial and actinomycetes counts. This alteration in the microbiological spectrum of soil may sometimes be harmful as the growth of useful bacteria and actinomycetes is impeded. Care should therefore, be taken while using ammonium fertilizers repeatedly on an annual basis and the dose of the fertilizers should be regulated so that it does not adversely affect the useful microorganisms of soil.

Temperature

Temperature plays an important role in regulating soil fungi. This aspect is of more concern in tropical soil where temperature variations in different seasons are quite pronounced. The fungi are mostly mesophilic and when the temperature reaches extreme conditions, it has an adverse effect on fungi. Except for a few thermophilic or psychrophylic species, high and low temperatures respectively are not conducive for soil fungi. Such extremes are observed in tropical and temperate situation. In tropical soils, the high temperature noted during hot summer month say April, May and a part of June, are unsuitable for the survival of fungi in soil. Mishra (1965) observed that in summer months the fungal population in soil decreased drastically but it suddenly increased just after the onset of rains in June and July. Normally, delicate hyphal structures, and hyaline thin walled spores of fungi die during summer mon.. ;, particularly on the upper surface of soil where the temperature recorded reaches up to 35-40°C. The effect is both directly through heat and indirectly through desiccation of soil due to a low moisture status. Certain fungal propagules like thick walled spores, rhizomorph, sclerotia etc. help in the survival of fungi under such an extreme situation and on account of them the perpetual survival of fungi in soil is achieved when the normal soil condition is attained. Certain fungi, however, exhibit wide tolerance for temperature. The author has isolated species of *Aspergillus, Penicillium* and *Trichoderma* from soil even during the summer months. It has also been observed by the author that in the hot months of summer, the sub layer of soil sometimes harbours more fungal populations due to favourable temperature and moisture regimes than the top layer of soil (Mishra, 1965).

Moisture regime

Population density of soil fungi is positively related to the moisture level of soil. Abundance and functions of fungi are directly affected by soil water content. Like other living organisms, in case of fungi too, water is the major constituent of living cells, 15-30% water content is quite conducive for soil fungi. Any deviation from this range is not favourable for most fungi. As is the case with other parameters, exceptions are observed with relation to moisture regime also. Certain soil fungi may grow above and below the said moisture regimes. An excessive moisture level in soil is harmful to fungi. This condition is not uncommon during the rainy season in our country, when water logging occurs in low lying areas. In places where water stagnates, the diffusion of O_2 necessary for aerobic metabolism is inadequate for microbiological demand and in such situations fungi are among the first to suffer. Most of the genera are detrimentally affected due to excessive moisture. Orpart and Curtis (1957) and Mishra (1965) observed that Mucorales in general are not affected much by excessive moisture.

Aeration

Fungi being aerobes are most abundant in the upper few inches (4") of soil where gaseous exchange takes place freely. At the lower horizons the population of fungi decreases due to poor O_2 availability and due to the accumulation of CO_2. Sometimes it has been observed that in the case of filamentous fungi, a part of the mycelium penetrates deeper into the soil, where O_2 level is poor or completely absent. In such a condition, the major portion of the mycelium is, however, in contact with the upper horizon of the soil where O_2 concentration is fairly high. In water logged condition, when free gaseous exchange is reduced, the population of fungi is also adversely affected and only the few species endowed with resistant spores might survive. Normally such water logged conditions are short-lived and as soon as the water is drained off, the surviving species recover rapidly and a normal situation is attained.

Profile

As has been repeatedly emphasized in the preceding pages the upper few inches of the soil profile are rich in organic matter, have free gaseous exchange and are adequately moist to support the better growth of fungi. Fungi are hence plenty in the upper horizon. In addition, roots widely spread in the region favour the fungal flora in many ways which will be discussed later in the chapter dealing with the rhizosphere. It has been noted that at deeper depths of soil, say after 40-50 cm. the fungal flora is almost negligible and this environment is suitable for the growth of certain anaerobic bacteria etc.

Season

The seasons of a year have a pronounced effect on soil mycoflora. Marked seasonal changes are observed in tropical places. Hot summer months as stated earlier, are not suitable for fungi and this is the time when the lowest fungal counts are reported. Varying temperate and moisture regimes coupled with nutrient availability in soil, which are closely linked with seasonal changes, affect the soil fungi. The seasonal cropping system has its own effect on fungi. In temperate conditions in the winter months when the temperature level drops down to a steep low, fungi find it difficult to survive. However, in such adverse conditions too, some fungi which are equipped with different types of resting bodies, survive with low biological activity and as soon as conditions become favourable the population status is regained.

Vegetation

Our discussion on the determinant factors regulating the population and activity of fungi in soil will not be complete unless attention is given to soil vegetation. The structure and property of the soil is a result of the cover vegetation on it. Plants growing in a locality change the nutrient status, moisture regime and aeration. There is a regular supply of plant debris to the soil surface, which enhances the nutrient status. With the addition of plant litter the activity of fungi in soil is enhanced. Depletion in litter accordingly decreases the fungal population. There are reports that certain fungi are specifically associated with particular plants. This picture, however, is not clear due to the use of conventional methods like dilution plate counting. As stated earlier, amendments of agricultural soil by manuring and fertilizing for better crops also have its own effect on the soil fungi.

Actinomycetes

Actinomycetes is a transitional group between bacteria and fungi and in soil they are numerically second to bacteria. Structurally they are unicellular organisms that produce slender branched mycelium and undergo fragmentation. They also produce asexual spores by the division of mycelium. Morphologically the branched, aerial, individual hyphae are similar to fungi but are less broad, with a diameter in the range of 0.5-1.2 μ. Some actinomycetes also produce by asexual spores known as conidia. Sexual stage in the life of actinomycetes is not known.

This is a peculiar group in that they exhibit characters of both bacteria and fungi. Alexander (1961) opined that, "despite their placement together with bacteria, the relation of the actinomycetes to the fungi, particularly the fungi imperfecti is apparent in four properties : (a) the mycelium of the higher actinomycetes has the extensive branching characteristic of the mould, (b) like the fungi many actinomycetes form an aerial mycelium as well as conidia, (c) the growth of actinomycetes in liquid culture rarely result in the turbidity associated with unicellular bacteria rather it occurs in distinct clumps or pellets and (d) the growth rate of atleast some strains is unrestricted and is not exponential as is the case with bacteria but cubic, a characteristic in common with many of the fungi". The actinomycetes resemble bacteria in morphology and size, conidia and the individual cells after the fragmentation of the mycelium. Certain actinomycetes are similar to *Mycobacterium* in general morphology, staining reaction and physiology. Like bacterial cells they are also susceptible to the attack of viruses, a feature not observed in fungi.

On account of their similarity to fungi in morphological appearance, the group was earlier called "ray fungi". In the light of recent information it is now apparently clear that actinomycetes have a closer association with bacteria than fungi. Besides morphological similarities actinomycetes possess nucleoids, a character exhibited by bacteria. Also, chitin or cellulose compounds characteristic of fungal cell walls are not found in actinomycetes, on the contrary, like Gram positive bacteria cell walls of actinomycetes are made up of polymers of sugars, amino sugars and a few amino acids (Cummins and Harris, 1958).

Actinomycetes as a group exhibit extreme variability. Morphological, cultural and physiological variations are very common among the actinomycetes. In normal conditions the mycelium consists of non-septate long hyphae, some of which are straight and quite lengthy, sometimes more than 600 μ others are much shorter, branched and curved. Branching is generally monopodial. The vegetative mycelium mostly is coloured, cream, yellow, orange, red, green, brown or black. Initially the cytoplasm of cells is quite homogenous which later on becomes vacuolated in an older stage of growth.

In certain genera like *Streptomyces* aerial mycelium is abundant and it covers the whole colony. When fully grown it gives a cottony or powdery appearance. Aerial mycelium may be fertile or sterile, a sterile mycelium is thin where as fertile ones are thinner initially but later on an increase in thickness is observed. Sporophores are arranged on the fertile part of aerial mycelium. The sporophores can be long or short, straight or practically curved. The arrangement of sporophores varies in different genera of actinomycetes.

Actinomycetes are widely distributed in different types of soils but they are more abundant in surface soil. In soil with a high pH the population of actinomycetes is very high. Composts, river muds and river beds are other places rich in this group. Physical characteristics, organic matter content and pH of soil determine the population of actinomycetes in soil. In dry soil with a high pH the population is much higher than other microorganisms. Their numbers are also considerably larger in grasslands and pasture soils than in cultivated fields. Virgin soils are relatively poor in actinomycetes. Environments with a pH less than 5, like peats and water-logged localities, are not conducive to the group.

The majority of actinomycetes are aerobic, except *Actinomyces* which are an aerobic or micro-aerophilic. Most actinomycetes are mesophilic, but forms like *Streptomyces*, *Thermoactinomyces* and *Thermomonospora* are thermophilic.

Depth, moisture content, soil reaction, soil type and soil vegetation as in the case of bacteria and fungi, are important factors which influence the occurrence and distribution of actinomycetes in soil. Waksman (1959) noted that the number of actinomycetes decreases with the depth of soil, but they increase in proportion to bacteria by 10-65%. Szabo *et al.* (1958, 59) were also of the same opinion, but they observed that sterile types were dominant in the deeper layer (B-horizon) while sporulating types were more frequent in the A horizon of soil. Better aeration and drier condition of the upper layers in their opinion were responsible for dominance of the group in A horizon.

Streptomyces are high salt tolerant and on account of this, different species of the genus occur frequently in saline soils and sea water (Grein and Meyers, 1958). A high population of the genus is also widely reported from cultivated fields which are better aerated than non-cultivated soils.

Alexander (1961) classified actinomycetes within the order Actinomycetales. The latter comprise four families and nine genera, only the last seven of which are included by the non-taxonomic term actinomycete.

I. Mycobacteriaceae

Mycelium rudimentary or absent. Spores absent.

A. *Mycobacterium* : Gram positive, aerobic, mesophilic rods commonly acid fast, non-motile and usually non-branching. Soil saprophytes are included in the genus.

B. *Mycococcus* : Gram positive, aerobic, mesophilic cocci, not acid fast. Found singly, in clumps or in short chains.

II. Actinomycetaceae

True mycelium formed. Spores produced by hyphal fragmentation. Spores not in sporangia. In early growth, the mycelium is continuous, but subsequently it fragments into bacillary or coccoid segments.

A. *Actinomyces* : Anaerobic or micro-aerophilic. conidia not formed. Typically parasitic, causing human and animal diseases.

B. *Nocardia* : Obligate aerobes. Cells occur as slender filaments. Aerial mycelium rarely formed and conidia not produced. Colonies similar to those of true bacteria, common soil organisms.

III. Streptomycetaceae.

True mycelium formed. Vegetative mycelium does not fragment into small segments. Spores produced but not in sporangia.

A. *Streptomyces* : Conidia formed in chain on aerial hyphae, aerobic. Abundant in soil.

B. *Micromonospora* : Conidia formed singly at the terminal end of short conidiophores, never in chains of spores. No growth at 50-65°C.

C. *Thermoactinomyces* : Similar to the Micromonospora except for growth at 50 to 65°C.

IV. Actinoplanaceae

True mycelium formed. Spores produced in sporangia.

A. *Actinoplanes* : Sporangiospores motile. Aerial mycelium rare.

B. *Streptosporangium* : Sporangiospores not motile. Aerial mycelium common.

Amongst the above genera *Actinomyces, Actinoplanes* and *Streptosporangium* are isolated from soil but they are not very common. *Micromonospora* and *Thermoactinomyces* are able to grow in soil but they are relatively much less in number than *Streptomyces* and *Nocardia*. *Streptomyces* is abundant in aerable soil, next comes *Nocardia*, and *Micromonospora* is generally scarce. According to the estimate of Alexander, 70-90% of the actinomycetes in virgin and cultivated fields are *Streptomyces*, less than a *third* are *Nocardia* strains and *Micromonospora* hardly contributes about 5 per cent. Some thermophilic forms occur in manure, compost (Hensen 1957) and mouldy hay (Corbaz *et al.*, 1963).

Like bacteria and fungi, actinomycetes are isolated from soil by conventional plate method. Most of the isolation techniques used for actinomycetes are the same as in the case of the other two types of soil microorganisms i.e. bacteria and fungi. However, certain selected media are generally used for the group and the incubation period is a little more than that of bacteria because of the slow growing nature of the group.

Activities in soil

Important functions which actinomycetes perform in soil are —

(a) Decomposition of organic matter content, (b) Plant pathogens. (c) Antibiosis and (d) Microbial equilibrium (Kuster, 1967).

(a) Decomposition of organic matter

Actinomycetes are one of the most active decomposers of various types of organic matter. Their ability to decompose cellulose and other polysac-charides is well known. In order of preference, water soluble car-bohydrates are most readily attacked followed by hemicellulose and cel-lulose. On account of an efficient enzymatic set up they decompose a wide variety of substances available in soil. Waksman and Diehm (1931) reported the decomposition of hemicellulose particularly mannas and xylase by different species of *Streptomyces*. Degradation of laminarin and alginates in sea-weeds by *Nocardia* has been observed by Chesters *et. al.,* (1956). Both these organisms are active decomposers of cellulose in soil. *Streptosporangium* has also been noted to decompose cellulose. Certain ac-tinomycetes, like *Actinoplanes* and *Streptomyces,* can decompose keratin due to the presence of keratinase enzyme in them. Kuster (1967) noted a great increase in the number of *Streptomyces* in soil rich in *keratin* containing material. There are reports that certain members of actinomycetes are capable of attacking even lignin rich materials (Waksman, 1959). *Strep-tomyces* species play an active role in the formation of humic acid, which though quite resistant to chemical and microbial attack, has been reported to be degraded by *Nocardia* (Kuster 1950, 52).

By and large, actinomycetes are heterotrophic in their food habit and their occurrence in an environment is regulated by the availability of organic substrates. Besides general carbon compounds, certain unusual organic molecules like paraffin, phenols, steroids and pyrimidines are also the substrates for degradation by certain species of *Nocardia*. Similarly, strains of *Micromonospora* are well documented to decompose chitin, cellulose, glucosides, pentosans and possibly lignin.

It has been observed that actinomycetes are poor competetors. against bacteria and fungi in the early stages of decomposition of plant and animal tissue. This is more true when the residues are rich in simple carbohydrates. Therefore, during the process of succession on residues, actinomycetes appear a little later than bacteria and fungi. However, the situation is entirely different when the residues have to be decomposed at a high temperature. This is quite common in the rotting and heating of green manures, hay, compost and animal manures. Under such conditions, the thermophilic actinomycetes are the most dominant and they play an active role in the decomposition. *Thermoactinomyces* and certain *Streptomycetes* have the competitive advantage in such a situation.

(b) **Plant pathogen**

Amongst the many actinomycetes present in soil only a few act as plant pathogens. The most common pathogen is *Streptomyces scabies* which is responsible for potato scab disease. *Streptomyces alni* is associated with *Alnus glutinosa* (alder) and the organism forms root nodules which are now considered to be associated with nitrogen fixation. This aspect has been dealt with in a separate chapter.

(c) **Antibiosis**

Some species of *Streptomyces* are capable of synthesizing antibiotics. On account of this unique characteristics, the *actinomycetes* group as a whole has attracted the attention of many workers. Almost three-fourth of the *Streptomycete* isolates may produce antibiotics. The amount of antibiotics being produced in soil is not of much significance, but in the laboratory, when the conditions are optimal large amounts of antibiotics are produced. In soil, the situation is not that conducive due to limited nutrient availability and competition amongst the soil microorganisms. Though the amount detected in soil is very small it can exert a localized inhibitory effect. This behaviour in the root region has a special significance.

After the discovery of antibiotics, continuous screening of the actinomycetes and fungi of soil is done to assess their antibiosis activity. Actinomycetes are responsible for the synthesis of certain important antibiotics like streptomycin, chloreamphenicol, chlorotetracyline, oxytetracyline and cycloheximide.

(d) **Microbial equilibrium**

Besides the production of antimicrobial metabolites, few species of *Streptomyces* liberate extra-cellular proteases which lyse bacteria (Born, 1952). This has importance in the microbiological equilibrium in soil. Generally, in natural soil, there is a well balanced and typical microflora. The various factors which characterize the particular soil, maintain the microbial equilibrium. A decrease in the population of actinomycetes, particularly of *Streptomyces*, has been reported in cultivated fields of barley by Rehm (1960). However, in most cases such changes are short-lived and temporary and the original equilibrium is soon restored. Similarly with amendment of soil by fertilizers and organic manures, such changes are noticed but sooner or later equilibrium is again achieved. In all such cases the role of actinomycetes is of equal importance.

ALGAE

In contrast to bacteria, fungi and actinomycetes which are generally heterotrophs, the algae group as a whole being mostly autotrophic, occurs predominantly on the surface of soil where light is freely available. Besides light, the availability of moisture is also an important factor for the distribution of algae in soil. It is a common sight during rainy seasons, to note a film of algae on the soil surface which gives it a green colouration. The population of algae, however, is much less than the other three groups of microorganisms discussed in the preceding pages. Like other microorganism, algae too are more dominant during rainy seasons when they grow and multiply luxuriantly on the soil surface. Later on when the moisture regime of the soil is depleted, the algae on the soil surface become less visible and only at such places where water is stagnant one can see algal patches.

Soil algae are normally unicellular forms but it is not uncommon to find filamentous strains when conditions are favourable. In general, they are smaller and less developed than their aquatic counterparts. As stated above, algae are the most abundant at or close to the soil surface but they are also isolated from the lower horizons. Many workers have reported a high density of algae below the surface of soil (Flint, 1958). Feher (1948) reported nearly 700 species at 15 to 20 cms below the soil surface. He further opined that this is the depth "where can be found the highest intensity of soil life". Movement of algae to lower horizons is a matter for discussion. Normally rain and soil animals particularly earthworms have been identified as agents for the downward movement of algae. There is always a tendency for such algal species to return to the surface whenever they can get the opportunity. This becomes difficult however, if they are buried too deeply.

The most common soil algae belong to chlorophyceae, cyanophyceae, bacillariophyceae and xanthophyceae. Nearly all the diatoms and cyanophyta are motile forms and many chlorophyta and xanthophyta produce zoospores.

Autotrophic species found on the surface of soil are independent of the preformed organic matter and like other green plants they manufacture their food, through photosynthesis. From soil the autotrophs need nitrogen, potassium, phosphorus, magnesium, sulfur, iron and other micro-nutrients. From the atmosphere they derive CO_2 and energy in the form of light, once all these requirements are met with, the autotrophic algae thrive in soil. The situation is different for those strains which are located below the surface where darkness prevails and photo-autotrophic life is difficult. It is interesting to note that although many algae are obligate photo-autotrophic it is not uncommon to observe heterotrophy in several species of chlorophyceae cyanophyceae and diatoms. For such heterotrophic strains, the energy source is not light but is rather the oxidation of organic carbon. Such algae known as facultative photo-autotroph are capable of metabolizing a large variety of carbohydrates like starch, sucrose, inulin, glucose, galactose, glycerol and citric acid.

There is a marked difference, however, in the dominance and activity of the algae formed at the upper horizons and those localized at deeper depths. At the surface, algae are favoured since they are not restricted by the organic matter level and also they have not to compete for organic carbon. On the other hand the algal species at lower depths, being heterotrophics, have to compete with other heterotrophic microorganisms which are well adapted to such a mode of life. It has been a matter of arguement whether subterranean forms have an active metabolism or they just survive passively in soil. The studies of many workers have thrown light on the subject. It is presumed that while a large number of algae move downwards through water seepage, tillage practices or by faunal agencies, certain species get adapted to new environments and they grow and multiply with in the horizon.

In most of the soil studies it has been noted that green algae, diatoms and blue green algae are abundant in soil. Members of xanthophyceae and flagellate chlorophyceae are less frequent. Geographically in temperate climates members of chlorophyceae are more frequent where as in tropical soils blue green algae are dominant. In virgin soil like those formed after volcanic eruptions or new islands emerged from the sea, the first colonizers on the soil are the members of algae and after the death and decay of algal cells, other microorganisms appear. Algae are ubiquitous in their distribution and are present in various types of soil.

Forms like *Chlamydomonas, Chlorella, Chlorococcum, Ulothrix, Vaucheria, Scenedesmus, Stichococcus, Pleurococcus, Protococcus, Hormidium, Cladophora, Coccomyxa* and *Dactylococcus* are certain dominant members of chlorophyceae commonly isolated from soil. Amongst diatoms, the important soil algae are *Cymbella, Navicula, Nitzschia, Pinnularia, Surirella* and *Synedra*. It is interesting to note that diatoms frequent in soil are characteristically smaller sized ones. It appears that the environmental conditions in terrestrial habitats favour development of smaller individuals. Small size has its own advantage because it permits greater water and salt absorption on account of the greater surface : volume ratios of the cells.

In addition to the above two groups, soil is very rich in cyanophycean algae. This group has attracted the attention of many researchers because of the unique character exhibited by certain members of blue green in fixing atmospheric nitrogen. In addition to chlorophyll and carotenoids, the group has a blue pigment known as phycocyanin which imparts the characteristic blue colour to cells. Cells lack a clearly defined organised nucleus like other green plants and in this respect blue green algae bear more of a resemblance to bacterial cells. The most frequently observed soil blue green algae are *Chroococcus, Cylin- drospermum, Lyngbya, Nodularia, Anabaena, Nostoc, Oscillatoria, Scytonema, Tolypothrix* and *Phormidium*. The group as a whole prefers neutral to alkaline soil. This character has wide implications and great economic importance. Many species of blue greens have been suggested to be useful in reclamation of alkaline soil, so called 'Usar' soil in India. In our country, a lot of research has been done in this aspect and based on these findings it has been indicated that simply by putting such alkaline soils under water coverage for a certain period and allowing the members of blue green algae to grow on wet soil for a few years, the soil may be made fertile.

Factors affecting soil algae

Important factors regulating the distribution of algae in soil are moisture, temperature, pH, inorganic nutrients and organic matter in case of heterotrophic algae.

Algae by and large being photo-autotrophs differ from other soil micro organisms in their requirements. Light and CO_2, the two major factors regulating their distribution, are generally not in short supply. Obtaining an adequate supply of CO_2 rarely poses a problem as CO_2 and bi-carbonates are usually produced in excess of the demand. Similarly light in normal course is also adequately available for autotrophs. However, light is a determining factor with respect to the vertical distribution of soil algae. At deeper depths light is a major determinant. As such the population is most dense in the upper 5 to 10 cm soil which falls off drastically with depth. However, as indicated earlier algae are present far below the zone of light penetration and in certain cases counts upto 10^3 per g have been recorded from C horizon. However, under such unfavourable condition the algal cells probably exist in a dormant state.

In contrast to other soil microorganism, algae as a whole prefer continuous saturated soil. Unless the soil becomes anaerobic due to long periods of water logging, prolonged saturation permits new communities to appear. Under excessive dried situation cells normally may die. However, like other microorganisms they survive by producing thick walled resting bodies which are resistant to dry condition. Members of cyanophyceae possess unique characters and have the ability to withstand very wet and very dry conditions and this accounts for the prevalence of the group in various types of soil (Fery and Fogg, 1962). They have the capacity to withstand rapid changes from one to the other. It has been observed that in dry period they exist as mats or crusts and on availability of sufficient moisture the cells rapidly imbibe water, expand and are not easily displaced.

Temperature too has its effect on the densities of algal cells. Though the algal cells are quite resistant to a wide range of temperature, the rapidity with which it rises and falls affects the soil algae. Drought may have an adverse effect and lead to the death of cells, similarly, low temperature is important in relation to loss of water (Hoffer, 1951). Reports regarding the abundance of algae in alpine, arctic and antarctic areas are available. Similarly, wide spread distribution of soil algae in hot climate has also been reported. Occurrence of cyanophyceae in severe climatic conditions is quite frequent. This is also a reason for the successful colonization of cyanophycean members in any extreme environment. The temperature range for algal distribution is quite wide being as low as 11.5°C to the higher range of temperature like 87°C (Kovada *et. al.* 1956).

As stated earlier algae prefer neutral or alkaline soil and so the widest diversities of species are found within this range. It has also been noticed that in acidic soil, members of chlorophyceae are more common whereas in alkaline environments cyanophyceae are abundant. Diatoms are less frequently encountered in acid soils whereas they are abundant in calcareous areas.

Amongst nutritional aspect carbon, nitrogen and phosphorus have been intensively investigated. Most of the information available, however, is based on the studies conducted under laboratory conditions. Besides the above nutrients calcicum (Ca) is another element which also plays an important role in algal distribution. The effect of Ca is more pronounced on the quality of the algal flora than in its richness. This is because calcareous soils are rich in nitrates and phosphates and hence they favour rich algal flora. The addition of fertilizers to agricultural soil, encourages the growth of cyanophyceae. Knapp and Lieth (1952) observed that a mixture of phosphorus and potassium is quite effective, where as ammonium nitrate is relatively less effective. Similarly potassium has less effect than phosphorus (Schwabe, 1963), Lund (1945-46, 47) in a series of studies suggested that effects of phosphorus and nitrogen are quite pronounced on algae and soils rich in phosphates and nitrates were also quite rich in algae. Diatoms, on the other hand, are less affected by phosphate (Shtina, 1956 b). Singh (1961) observed an increase in cyanophyceae in rich fields when phosphate was added.

An interesting interrelationship has been observed between soil algae and other microorganisms of soil. It has been noticed that the growth of heterotrophic algae in dark is promoted by bacteria and actinomycetes found in the region. Stimulatory effect of certain cyanophyceae members on *Azotobacter* has been reported (Federov 1952, Shtina and Yung 1963). This is particularly true when *Nostoc* and *Azotobacter* are grown together. Interaction has also been reported between algae and fungi or bacteria. Fletcher and Martin (1948) observed that if the soil is kept moist for more than a week, the algae denatured and fungi took their place. Parker and Bold (1961) observed an increased growth of a coccoid green alga in the vicinity of a bacterium. This probably was on account of the ability of the bacterium to decompose organic nitrogen compounds, which in turn proved useful to the alga. Parker and Turner (1961) suggested that the majority of the heterotrophic organisms, with the exception of protozoa, stimulated the growth of algae. Surprisingly in many cases the growth of a fungus may depend on the presence of sufficient algal cells in its neighbourhood.

Interrelationship is not only confined amongst the microorganisms. Algae are known to influence the growth of higher plants as well. Such effects may be of various nature i.e. inhibiting or stimulating to one another. Shtina (1954) observed that germination of seeds may be unaffected, stimulated or inhibited by algae. She also produced evidence that roots of angiosperms as such are not affected by soil algae but algal extracts may stimulate their development. On the contrary, root of certain plants may stimulate algal growth. Inhibition in the growth of phytopathogenic fungi by algal extract has also been noticed.

The algal population in soil plays an important role in soil fertility. In places like rice fields, where the population density of algal cells is very high, the dead or live algae may act as fertilizers. In deserts and other barren soils, algal crusts may prove useful in binding soil particles together. This has great ecological importance because it helps in prevention of soil erosion and facilitates recolonization of the soil by higher plants.

As indicated earlier the most important function of soil algae, particularly of certain members of cyanophyceae, is to fix atmospheric nitrogen. A large number of cyanophycean members can fix atmospheric nitrogen symbiotically or non-symbiotically. This aspect will be dealt with in a separate chapter.

Protozoa

The phylium protozoa is represented by primitive, unicellular organisms, which vary in size from several microns to one or more centimeters. Though the microorganisms discussed earlier constitute the major population of soil and are categorised in microflora, protozoa being representatives of the animal kingdom are in no way less important in soil ecosystem. Along with soil microflora, this group plays an important part in the ecological and biochemical processes operating in soil. Normally cells of protozoa are devoid of chlorophyll, but interestingly there are a few transitional genera which resemble algae and are provided with chlorophyll pigments. Furthermore, there are certain genera which are claimed both by plant and animal experts as being in their domain. The animal representatives grouped in protozoa exhibit the simplest form of animal life.

Two prominent stages are encountered in the life cycle of protozoa - an active phase in which the organism feeds, multiplies and leads an active life and a resting or cyst stage. The latter stage is usually observed when conditions are unfavourable. The cell then secretes a thick covering around itself and forms a cyst. Resting or cyst stage helps the organisms to withstand unfavourable conditions and on the return of normal conditions the cyst germinates to give rise to an active cell. The normal mode of reproduction in protozoa is asexual. The mother cell divides longitudinally or transversally into two halves which are called daughter cells. In certain protozoans sexual reproduction has also been reported. The process of sexual reproduction is very simple; morphologically similar mating cells fuse and exchange of genetic materials takes place between the two. After the exchange, the two cells separate and new individuals emerge.

Protozoa are of wide occurrence in different types of soil. Like other soil organisms they are also recorded in almost all sorts of soil. Variation in the density and composition of protozoa is quite common in different soil types. Certain soils may contain a few types only whereas others may exhibit a wide variety of them. Similar ecological niches harbour different faunas as a result of accidents of zoogeography.

Protozoan species are generally defined on morphological grounds. Mode of locomotion forms the basis of classification of protozoa. Some protozoa are provided with one or more long flagella, others have short hair - like cilia and there is a third group where movement is by means of temporary organelles known as pseudopodia. Some of the parasitic genera on the other hand lack specialized structures for movement. The following are the five major classes identified in the phyllum protozoa.

(a)   Mastigophora or flagellates - locomotion by means of flagella (1-4 number)

(b)   Sarcodina (= Rhizopods) - locomotion by means of pseudopodia

(c)   Ciliata or ciliates - cilia present throughout the active phase of life cycle

(d)   Sactoria - cilia present only during young stage

(e)   Sporozoa - no specialized locomotory organelles.

In soil only the first three types of protozoa are found and hence the following discussion will be based on them only.

(i)   Phytomastigophora and

(ii)   Zoomastigophora

Phytomastigophora are usually endowed with chlorophyll and like normal plant cells they grow photosynthetically. The forms included in this group, as indicated earlier are claimed both by plant and animal scientists. Representative of zoomastigophora are typically animal like and are devoid of chlorophyll. The organisms included in mastigophora are motile and they possess normally 1 to 4 agella. In exceptional cases, in certain species the number may be more than four also. *Euglena* and *Chlmydomonas* are the best examples of phytomastigophora. Besides the two other representatives common to soil are *Cercobodo, Eutosiphon, Monas, Spiromonas, Spongomonas* and *Tetramitus*.

In contrast to mastigophora, the members of *Sarcodina* do not possess permanent locomotory organs. Movement in this case is by means of temporary protoplasmic extensions from the body of the cell. The unique feature of the cell in *Sarcodina* is the lack of a rigid external surface and this remits frequent change in the animal body. The organism, as and when required, produces protoplasmic extrusions from the cell body, which are known as pseudopodia. Pseudopodia as indicated are not permanent structures, but simply as extension of the protoplasm and are sent forth or withdrawn as the need of the organisms may be. Certain species of Sarcodina may be provided with shell like structures and in such cases the pseudopodia extend through distinct openings in the shell. *Amoeba,* Diffiugia, *Euglyphia, Naegaeria, Hortmanella* and *Trienma* are typical representative of this class.

In ciliata there are numerous hairs around the cell and movement of the organism is through the action of these hairs. In some species several thousand hairs may be found on a single cell. Ciliates are normally very small in size, ranging from 10 - 80 μ in length. Some typical soil ciliates are *Colpoda, Vorticella, Oxytrichia, Colpidium, Halteria* and *Balantiophorus*. A few soil protozoa are polymorphic. Most of the flagellates are monomorphic but some, like *Amoebae,* occur in either flagellated or limax form. This change is induced by altered environmental conditions (Willmer, 1956). Amoeboid organisms such as the *Proteomyxid* have both limax and a reticulate stage. Ciliates such as *Tetrahymena rostrata* have theront, trophnot and tomite stages (Stout, 1954) and free swimming and tentaculiferous forms are found in few soil suctorians.

Soil protozoans are characterised by the presence of cysts or other forms which

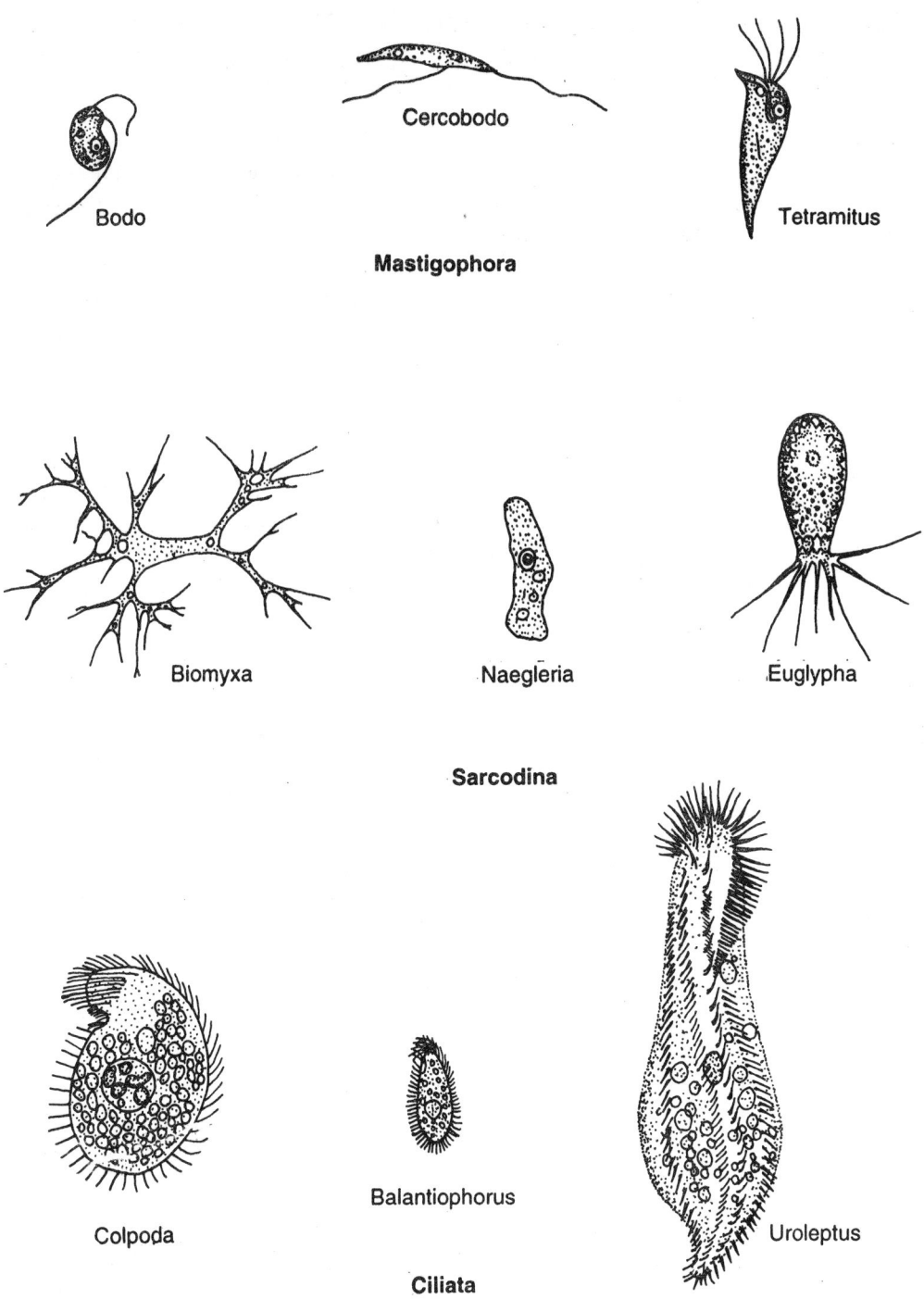

Fig. 1.1(d) Common protozoa in Soil

help them in resisting desiccation. Cysts are of variable nature in different protozoans. In Amoebae they are simple, typically spherical with varying structures and thickness of wall (Singh, 1952). Normally starvation induces cyst formation, however, other factors like decrease in aeration and divalent salts could cause encystation. Oxygen supply also limits cyst formation in a few protozoan species like *Hartmannella rhysodes*. Band (1959) observed that in this genus no cysts are formed at low O_2 tension. Excystment in some species may be obtained in water, whereas in few cases presence of bacteria or their metabolic products like amino acid may be helpful. Goodey (1915) observed that cyst of flagellates and amoebae remain viable for a long period in soil. Volz (1929) noted that in testacea several types of cysts are formed. Bonnet (1964) observed that physical and physiological drought appear to be important factors for encystment and excystment. In testacea the life cycle in general is very complex and this has not been properly understood.

In ciliates too cystment is of varying nature. In *Colpoda steinii*, the most wide spread soil ciliate, three types of cysts have been observed - resting or resistant cyst, reproductive cysts and unstable cysts. Exhaustion of food and crowding induce resting or resistant cysts. Reproductive cysts are associated with growth of the cell and after attaining a certain stage, such cysts are formed within which cell division takes place. Unstable cysts are formed in response to adverse conditions and in this situation growth and cell division are inhibited.

Method of study of protozoa from soil

Three methods are generally employed for the study :
(a) Direct observation
(b) Extraction and
(c) Culture

(a) **Direct Observation**

As stated in case of bacteria and fungi the presence of protozoa is soil in examined by inserting slides (Storkey, 1938) or by examining the soil as such, in water to which a stain has been added (Volz, 1951). By this technique, testacea are more readily recognized than other groups because their hard testa can withstand the rough treatment of preparation.

(b) **Extraction**

Carbon dioxide or air is bubbled through the soil and the tests containing gas float to the surface (Chardez, 1959). Extraction of ciliates using an electric current has also been successfully tried by Horvath (1949) and Hairston (1965). This method in general is more suitable for testacea.

(c) **Culture**

Culture techniques are the most satisfacotry for qualitative studies. Dilution culture technique has been extensively used for estimating populations. Media enriched with peptone or yeast extract are preferred over non-enriched ones. As the soil protozoa are generally bacteria feeders addition of bacteria to cultures encourages the growth of protozoa.

As with other soil microorganisms, with protozoa too one method is not sufficient to give a complete picture of the protozoa in soil and it is desirable to employ more than one technique for better assessment.

Mode of nutrition

Various modes of nutrition are reported in soil protozoans. Some important one are the following:

(a) Photoautotrophy

(b) Saprozoic

(c) Holozoic

(a) **Photoautotrophy** - certain chorophyllous phytoflagellates like *Euglena* and *Chlamydomonas* are capable of synthesizing protoplasm from CO_2 using energy obtained from sunlight. Though it is a typical plant behaviour and is generally not an attribute of animal cells, alga - like flagellates of the protozoan group are endowed with this advantage and as stated earlier they are a connecting link between animals and plants.

(b) **Saprozoic** - Most protozoa are dependent on organic matter for their nutritional requirement and they derive their nutrients from organic and inorganic substances. In this respect they are very similar to other soil microorganisms.

(c) **Holozoic** - A vast majority of protozoa are holozoic and they feed directly on microbial cells. Normally such protozoa feed on bacterial cells. Once the food material is ingested, it is surrounded by a vacuole wherein digestion takes place. Such materials are rich in proteins, polysaccharides, sugars and toprids and protozoa have the ability to mediate decomposition through suitable enzymes. Ultimately, the undigested part of food is released outside. Predation on the bacterial flora is selective. Different strains may vary as much as different species in suitability (Hetherington, 1933; Leslie, 1940, a, b). Gram-negative Enterobacte- riaceae are the most favoured one and the Bacillaceae the lowest (Burbanck, 1942). Using many bacterial strains and a wide range of soil flagellates, amoebae and ciliates, Singh *et. al.*, (1958) classified bacteria into 4 categories:

(1) Strains readily and completely eaten

(2) Strains slowly but completely eaten

(3) Strains partly eaten and

(4) inedible strains

Edibility of bacteria by Amoebae is also related to pigmentation of the bacterial cells. Singh (1945) observed that pigments present in the cells of certain inedible strains are toxic to the protozoa. *Pseudomonas aeruginosa* and *P. fluorescens* are reported to be toxic. Graf (1958) observed that extra-cellular substances derived from *P. fluorescens* contain B amino acids and a fatty acid. Amoebae grow but with sarcinae, cocii and Gram-negative rods; poorer growth occurs with bacilli and diphtheroid. Groscop and Brent (1964) were of the opinion that the size of bacterial cells and physical characteristics of their surfaces are more important than chemical differences such as pigment production in the selection of food by amoebae. Also the age of the bacterial culture is another important aspect for selection. *Agrobacterium, Bacillus, Micrococcus* and *Pseudomonas* are good food source to protozoa.

Actinomycetales in general are not suitable as food for amoebae and possibly have antagonistic effects. Exudates from soil *Streptomyces* are known to be toxic to soil amoebae.

Similarly protozoa very rarely feed on spores and hyphae of fungi. some species of protozoa have, however, been reported to use hyphae as food (Hervath, 1949). Yeasts have been reported to be eaten by protozoa (Hedrick and Traver, 1964). Heal (1963 a) observed that 19 species of yeasts were consumed as food by atleast one of the four species of amoebae. It thus appears that after bacteria, yeasts whose population in soil is reasonably high, $(10^3-10^{+6}/g)$ soil may form a staple food material for soil protozoa.

There are reports that few protozoa are obligate algal feeders and the ciliate genus *Nassula* and some rhizopods feed on algal cells.

In the absence of edible bacteria or when there is paucity of food material in the environment, the active protozoan enters the cyst stage. Besides food, low O_2 tensions or other types of unsuitable environments are other factors leading to encystment. Such bodies are more resistant to adverse conditions particularly to harmful chemicals, acid, germicides and high temperature. With the help of cyst, protozoa survive unfavourable conditions, however, rejuvenation of all the cysts on the return of favourable conditions is not guaranteed. Generally, it has been noticed that excystment is quicker if bacterial cells are in the vicinity of the protozoans. The effect is, however, preferential; some bacteria allow rapid excystment, other favour a slow return and a few though, allow quick but incomplete excystment.

Amongst the various factors affecting distribution and density of protozoa in soil, the most important are the presence of adequate food supply and moisture. The population of protozoa is the highest near the surface of the soil, particularly in the upper 6 inches. At lower depths the population is generally low. This is directly related to bacterial flora in the region. In the upper layer of soil, more numerous bacterial species account for the higher protozoan cells. As a corollary, any soil amendment which favours the growth of bacteria in soil, also helps in the rapid built up of the protozoan population. In agriculture fields, where manuring and fertilizing is done, both bacteria and protozoa are abundant.

The moisture level of soil affects protozoa both in quality and quantity. As for the other organisms, for protozoans too, adequate water supply is essential for physiological activity. Water also helps in the lateral and vertical movement of protozoa. It has been noticed that flagellates are tolerant to low moisture and they can develop in drier conditions. On the other hand, ciliates are moisture loving animals and they are abundant only when the moisture level is high.

Generally protozoans are aerobic and hence a steady supply of O_2 is necessary for their good growth. Very rarely some species grow at low partial pressures of O_2 or under complete anaerobiosis. Such situations are short lived ones and as soon as sufficient O_2 is available, the cells resume their normal metabolic activity.

Most protozoa are indifferent to pH variations. Many species, however, are susceptible to extreme variations. Cool and damp conditions are favourable to protozoans. All these variables act together rendering it difficult at times to assess the effect of an individual factor on any population. For protozoans, however, as stated earlier any factor adversely affecting the bacterial flora has an adverse effect on them also. On the whole, the population, both of bacteria and of the associated protozoan fauna, is determined by other soil conditions - the nature of the organic cycle, the moisture status and the base status.

Part played by protozoa in soil

The role played by soil protozoa is not properly understood. There are conflicting reports available in this respect. Normally it is expected that many, protozoa, being bacteria feeders, regulate the size of the bacterial population. It was also suggested that, protozoa being detrimental to the growth of useful bacteria may affect soil fertility. Experimental evidences however, do not support such hypothesis. Protozoa mainly affect the organic cycle indirectly. Their constant predation on bacteria contributes to the general turnover of readily available nutrients. This appears to favour biochemical activity. An increase in the growth of higher plants, when treated with amoebae and ciliates, has been reported ((Nikoljuk, 1956, 65). Suppression in the growth of *Rhizoctonia solani* and *Verticillium dalliage* two phytopathogenic fungi by protozoa has also been observed.

Viruses

In earlier pages, the microorganisms described and discussed are small and form a substantial living population of soil. In addition to those, which are visible by light microscopy, an unique group of organisms are also found in soil, which unlike the other microorganisms does not form a major soil microbial population and is beyond the resolution of light microscopy but can be easily studied with the help of recent techniques like SEM and TEM. Such submiscroscopic organisms are called viruses and though numerically lesser in number in soil , play an important role and are economically very important. These infective agents are responsible for many diseases of plants and animals. There are many strains of viruses which also infect microorganisms like bacteria and actinomycetes and in such condition they are designated as bacteriophages and actinophages. Some viruses also parasitize cyanophy-ceal members and then they are known as cyanophages.

The characteristic feature of viruses is their intimate relation with a suitable host. Also, viruses are limited in their host range and as such they infect and parasitize only specific plants, animals or microorganisms. Viruses normally grow and multiply only when they are inside the host body. They do survive outside the host for varying periods but no activity is noticed during the period. Certain plant viruses like those responsible for mosaic disease of wheat, oats and tobacco and big vein disease of lettuce, persist in soil when the respective host plants are harvested. In a true sense, such viruses are not a native resident of soil and they simply live in the soil and retain their infective capacity as long as the host crop is not available. The period thus spent in soil may be for a year or little more. There are reports that certain animal and human viruses also survive in soil and retain their infective ability for some time. By and large the soil is more rich in bacterial and actinomycetes viruses i.e. bacteriophages and actinophages. Morphologically bacteriophages possess head and tail like structures. The diameter of the bacteriophage normally does not exceed 0.05 to 0.10 μ and the tail which is somewhat longer and quite narrow measures approximately 0.2 μ in length.

Viruses, as stated above, being submicroscopic are recognized by the symptoms they induce in the host. However, this recognition is possible only for macroscopic hosts which can be seen with the naked eyes. In case of microorganisms it is not easy to see the symptoms even under light microscopes. Indirect tests are adopted for the detection of bacterial and actinomycetes viruses. This can be done by growing the host on suitable nutrient media, on solid media the appearance of plaques or in liquid media the clearing of turbid cell suspension are indications of the presence of viruses within the inoculated host. The phages may then be spotted and later inoculated on cultures of susceptible host and allowed to grow. The phage particles can then be separated and purified by filtration and high speed centrifugation. Following this, the presence of bacteriophages can be demonstrated in a soil sample. A sample of soil may be inoculated with the host bacterium and incubated for 24-48 hrs. A small quantity of the treated sample is then added to a nutrient medium previously inoculated with the host and incubated again. The lysis is then clearly observed in the medium of the host cells. The suspension is then passed through a sterile bacteriological filter and the filtrate tested for its capacity to lyse a fresh growing microorganism. Using the technique bacteriophages of *Agrobacterium*, *Pseudomonas*, *Rhizobium*, *Streptomyces*, *Azotobacter* and *Nocordia* have been demonstrated in soil.

In soil, the bacteriophages infect bacterial cells and through the tail end, the content of the former is injected into the latter. Inside the bacterial cells the injected particle of the bacteriophage multiplies rapidly and produce a number of daughter cells. Such a type of bacteriophage is known as lytic or virulent. In most cases the daughter cells of bacteriophage are released after the death of the host cell and they form the fresh stock for further infection of healthy bacterial cells and the process continues resulting in the destruction of a large number of bacteria. Sometimes, however, the daughter bacteriophage cells are not immediately released out side the bacteria, rather they are retained with in the latter for varying periods. Infected bacterial cells, in such condition apparently look healthy and they continue to multiply and produce infected daughter bacterial cells. The bacterial cells do not lyse and occasionally release a few bacteriophages. Such bacteriophages are temperate and the phenomenon is known as lysogenicity and the bacterial cells which carry the bacteriophages are marked as being lysogenic bacterium. Lysogenicity has a special significance and is responsible for transmitting the genetic materials from one host to a newly infected bacterium. The phenomenon is known as transduction. In transduction an unidirection transfer of a portion of nuclear apparatus of microbial architecture takes place. Normally, such a transfer is achieved between strains of the same species, however, occasionally it may also take place between closely related genera.

The presence of bacteriophages in soil is of special significance in relation to certain useful bacteria, particularly those which are responsible for nitrogen fixation. It is quite possible that the excessively high destruction of such useful organisms on account of bacteriophage infection, may adversely affect nitrogen fixation which in turn may bring down the soil fertility level. However, normally the soil system is very stable and the minor fluctuation in the set up is a short lived one and sooner or later microbial balance is achieved. Viruses are generally adsorbed to organic residues, humus and clay particles. Soil rich in the above constituents are , therefore, rich in viruses too. On the contrary, sandy soils are not favourable for viruses and are poor in virus population. Taking this fact into considertion, the amendment of clay and humus rich soil with sand may reduce the virus population in soil.

The added advantage of microorganisms is their ability to go for genetic transformations at a rapid rate. Processes like transductions etc. many a time result in partial changes in the genetic setup of the daughter cells, which not surprisingly may be better equipped for various metabolic activities than the parent. The daughter cells may also prove to be more resistant to various pathogens, including the bacteriophages. The receptor cells may acquire additional physiological properties which might offer a competitive advantage.

Chapter 2

Microorganisms associated with plant roots

Soil microorganisms exhibit a range of relationships with the roots of higher plants. Some of the important ones are :

(1) Symbiotic - in the form of mycorrhizal associations and legume nodules

(2) Parasitic - in which case the causal organisms range from unspecialized to highly specialized forms and

(3) Less clearly defined relationship clumped together as rhizosphere and root surface phenomena.

As discussed in Chapter 1, soil is abundantly rich in various types of microorganisms which vary in form, structure and function. Soil is also an environment in which the underground parts of the plant have their abode. Roots of the higher plants are concentrated in the soil. As they grow through soil, the condition of the soil in the immediate vicinity of the root witnesses a drastic change in various ways. The microhabitat of the soil is in close proximity of root changes and this has an impact on the soil microorganisms residing in the region. Plants as such do not have a special excretory system and many substances in the form of waste products are released from different parts of the plant body. The root surface is one such region from where the unwanted substances from the plants are leached-out continuously and they get accumulated on the immediate surface of the roots. Substances thus released are known as root exudates. The region, just behind the root cap particularly, is the site of maximum exudation. Organic and inorganic substances exuded from the root enhance the microbial population in the region. In addition to exudates, sloughed off root cells, mainly derived from young growing root caps provide additional energy for microbial development.

The specialized region of the soil around the root which is influenced by root exudates is known as rhizosphere and this will form the theme of the discussion in the present Chapter. The term rhizosphere was coined long back by Hiltner (1904) and he defined the zone of enhanced microbial development around roots as the rhizosphere. Detailed studies on this aspect, however, were undertaken only after half a century of the identification of the region by Starkey (1929, a,b,c). Lateron many more researchers devoted their attention to this fascinating region and the classical findings of Starkey (1931), Katznelson (1946), Henderson and

Katznelson (1961) added to the knowledge in this field. With an increase in knowledge, various terminologies were introduced by various workers. Terms such as "histosphere", "rhizosphere" and "edaphosphere" (Perotti, 1926), inner and outer rhizosphere (Graf, 1950; Poschenreider, 1930) are the examples. Another term rhizoplane was also suggested by Clark (1949) to denote the external surfaces of plant roots and closely adhering particles of mineral soil and organic debris. There has been a lot of confusion regarding the terminology to be used for the region of enhanced microbial activity around the root. Rhizoplane, though used by many American and Canadian workers has not been widely accepted and most of the workers in the area prefer to use rhizosphere.

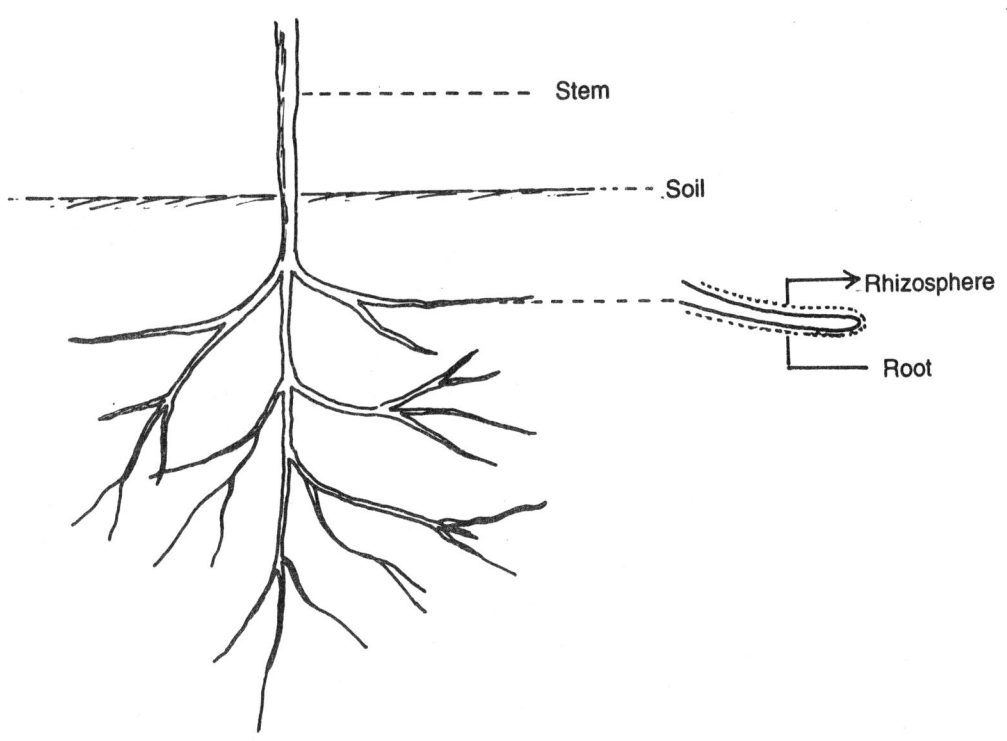

Fig. 2.1(a) Rost showing rhizosphere region

The rhizosphere region is very narrow in width and varies from plant to plant. Apart from this, even for the same plant the measurable rhizosphere effects are not the same through out the life cycle. Depending upon the age of the plant, the nature of root exudates differ and as a result the rhizosphere effect also varies. The rhizosphere effects have been measured in a few plants. In tomato roots the distance from the root is upto 5 mm (Rovira, 1953), upto 16 mm for lupin roots (Papavizas and Davey 1961). Besides age, the metabolic state of the plant and the nature of the soil also affect the distance. However, it has generally been noticed that the poorer the soil the more pronounced is the rhizosphere effect.

The results of most of the studies demonstrate that both number and activity of microorganisms in the rhizosphere attain peak at the time of maximum vegetative development of the plants (Mishra, 1967).

Methods for the study of rhizosphere microflora

Numerous methods have been suggested by various workers for the study of microorganisms in the rhizosphere region. Some important and conventionally adopted are as follows :

    (a)   Dilution plate method

    (b)   Rossi-cholodny buried slide technique

    (c)   Direct observation and

    (d)   Impression slide technique

(a)   Dilution plate method

The soil dilution plate technique has been widely used for the assessment of rhizosphere microflora (Starkey, 1929; Papavizas and Davey, 1961). Root samples are carefully collected and the extra soil attached to the roots is removed. The closely adhering soil is then used for the preparation of a soil suspension which is then diluted serially to obtain suitable concentrations of the suspension. The suspension is subsequently inoculated on convenient nutrient plates. The inoculated plates are finally incubated for varying periods depending upon the nature of the microorganisms to be studied. For bacteria the incubation period is normally 24-48 hrs; whereas for fungi and actinomycetes the period is longer, about 6-8 days.

Though this technique as stated above is widely used by the researchers for the isolation and enumeration of the rhizosphere microflora, it has been criticised for its lacunae. The technique is very selective and allows the isolation of a small proportion of microorganisms. It favours the growth of certain organisms which are heavily sporing and thus provides the sporing capacity of the microorganisms rather than the actual population present in the rhizosphere. It has been noticed that in the rhizosphere region the fungi are mainly present as mycelium whereas in soil away from the region they are found in the form of spores. As such any conclusion based on data collected by dilution plate method about the fungal population in rhizosphere and non-rhizosphere region has no relevance.

(b)   Rossi-Cholodny buried slide technique

Slides are buried in the soil in such a way that they come to lie in close contact with the roots of the plants. The slides are then left for sometimes to allow the rhizosphere microorganisms to grow upon them and are then taken out carefully so that the organism on the slides are not disturbed and dislodged. Proper stains are used and the organisms identified. This has some advantage over the dilution plate method in that the presence of the microorganisms *in situ* is recorded. However, it is extremely difficult to identify all the microorganisms seen on the stained Rossi-cholodny slides. Also, inspite of all the care being taken there is a great possibility of dislodging a few organisms from the slides during the operation. Through this method, however, the presence of clusters of bacteria in the root region and an abundance of fungal mycelium too have been demonstrated.

(c) **Direct observation**

Linford (1940, 42) used glass observation boxes for the direct observation of rhizosphere microflora. Krassilinkov (1958) described a method in which plants were grown on glass plates which are mounted in such a way that the roots spread over the glass leave their "imprints" on the glass. Microscopic slides are placed on the inner surface of the glass plate to facilitate the microscopic observation. These are removed at regular intervals for observation. Profuse microbial development on the surface of roots, between root hairs and some distance away from them can be demonstrated by this technique.

(d) **Impression slide technique**

The technique was designed by Brown (1958) and extensively used by Parkinson (1958). Microscopic slides are thinly coated with adhesive material like nitrocellulose in amylacetate and are pressed against freshly collected root samples containing adhering rhizosphere soil. After sometime, when the rhizosphere soil gets stuck to the slides they are removed carefully, stained and examined for rhizosphere microorganisms. The difficulty faced with this technique as in the case of the Rossi-Cholodny method, is the proper identification of the organisms. Furthermore, the amount of soil screened for the microorganisms is not definitely known.

Microbial population in the rhizosphere : A vast microbial population has been noticed on the surfaces of roots, root hairs and in their vicinity. Bacteria are generally localized in colonies and chains of individual cells. Filamentous fungi and actinomycetes though frequently observed in the rhizosphere are not that numerous. Among the protozoan, flagellates and large ciliates are conspicuously present in the water films on root hairs and on the epidermal tissue.

It has generally been noted that Gram-negative, non-spore- forming bacteria are stimulated to develop in rhizosphere soil. *Agrobacterium radiobacter* and *Pseudomonas* have been found to occur abundantly. In fact, it has been reported that the latter constitute 40-50% of the bacterial population of some rhizosphere (Rouatt and katznelson, 1961). *Mycobacteria* and *Corynebacteria* are other important genera of many rhizospheres. Most of the studies in the rhizosphere indicate that motile forms, chromagenic forms, ammonifiers, denitrifiers, gelatin liquefiers, forms giving an acid or alkaline reaction with glucose-peptone media and aerobic, cellulose decomposing forms are generally more in number in rhizospheres than in general soil. On the other hand, nitrifying organisms, anaerobic cellulose decomposing forms and nitrogen fixing anaerobes are less frequent in rhizospheres. Besides these, species of *Arthrobacter, Mycoplana, Brevibacterium, Flavobacterium, Serratia, Sarcina, Bacillus* and *Mycobacterium* have also been reported to occur in the rhizosphere of many plants. Among the Bacillus species, *B. circulans, B. brevis* and *B. polymyxa* are generally more common on the plant roots than outside the rhizosphere region (Clark and Smith, 1949).

In the rhizosphere region the microbiological competition is of a very high degree. This is on account of the high density of bacteria which are frequently as high as 10^9 cells per g. As a result of the high bacterial population and the stress resulting there-of in the rhizosphere zone, the fast growing organisms are normally favoured and they compete more successfully. It also appears that the biochemically more active species are favoured in the region and consequently, in the rhizosphere, the biochemical activities are more high than the non-rhizosphere region.

Different groups of bacteria have varying nutritional requirements and this accounts for specificity in the occurrence of bacterial species in the rhizospheres. As stated earlier, in the rhizosphere region the major nutritional supply is in the form of root exudates. The exudates are rich in a number of amino acids. Glutamic acid, asperatic acid, proline, leucine, alanine, cysteine, glycine, lysine and phenylanine have been reported in the exudates of different plants. In the presence of rich amounts of amino acid in the rhizosphere, the selection of bacteria in the region is associated with the quality and quantity of the amino acid. Lochhead and Rouatt (1955) observed that amino acid requiring bacteria make up a higher population of rhizosphere microflora than of the general soil microflora. They also remarked that the rhizosphere region is rich in bacterial species with simple nutritional requirements and a lower proportion of bacteria requiring yeast extract, than the soil distant from plant roots.

It will be appropriate to mention specifically the distribution of *Rhizobium* and *Azotobacter*, two nitrogen fixing species, in the rhizosphere region. Based on the studies of Clark (1948,49) it is evident that the rhizosphere as such is not a favoured environment for *Azotobacter*. Katznelson and Strzelczyk (1961) further confirmed that for 17 crop plants the *Azotobacter* density was very low not only in rhizospheres but also in their corresponding non-rhizosphere regions. Strzelczyk (1961) explained the low population of the bacterium in the rhizosphere. He showed that the rhizosphere in general was rich in a number of microorganisms which were antagonistic to *Azotobacter*, such antagonistic forms were, comparatively less in non-rhizosphere region.

As far as fungi are concerned roots do not alter appreciably the total counts of fungi. The influence on fungi is rather selective and a few genera are stimulated. The dominating genera in the rhizosphere of different plants vary with respect to the cover vegetation, soil and climate. Certain genera like *Aspergillus*, Penicillium, *Fusarium, Cladosporium, Trichoderma, Rhizopus, Mucor* and *Chaetomium* are more frequent. Besides these, a large number of other fungi also occur sporadically in the rhizosphere. *Trichoderma* is usually more dominant in acid or neutral soils, in alkaline soil on the other hand, *Penicillium* species are more abundant. *Rhizoctonia* and many unidentified sterile forms are also frequently isolated from rhizospheres. It has been observed that the rhizosphere mycoflora changes with an increase in age of the roots. In general, it has been noticed that during the young age of plants, phycomycetous species, particularly members of the *Mucoraceae* along with *Penicillium* species are more frequent. Later on when the roots become aged, this population has been shown to change, and then genera of Tuberculariaceae, Dematiaceae and sterile dark fungi are more commonly isolated. Chesters and Parkinson (1959) explained this change in the light of the type of nutritional material available in the root region at different ages of the root. They opined that in young roots the root exudates play an important role in regulating the mycoflora and at this stage sugar fungi are favoured, whereas in the rhizosphere of older roots, dead root material is an important cause of rhizosphere effect.

Different regions of the same root vary in their nutritional set up and the micro habitat as such in the root region is not uniform. Taking this fact into consideration, the rhizosphere environment will vary from point to point on a root. Samples from specific regions of the rhizosphere, that is from the root tip, crown of the root and a zone intermediate to the tip and crown, may exhibit varied types of mycoflora. Normally the change in the qualitative nature of the rhizosphere mycoflora with changing root age, begins first in the crown zone and lasts in the tip zone of roots (Chester and Parkinson, 1959).

Actinomycetes, protozoa and algae are not markedly affected by the rhizosphere effects and the rhizosphere : soil (R:S) ratios rarely exceed 2 or 3:1.

Stenton (1958) demonstrated that as young roots grow through the soil, they are almost free of microorganisms and they present a 'virgin ecological niche' for soil microorganisms. Before the primary root emerges, substances rich in amino acids are liberated from the seed and this continues from the root after its emergence. Jackson (1960) demonstrated the role of such substances. He stated that such substances help in releasing fungal spores from fungistatic operating in the soil and then allowing their germination and subsequent growth along a diffusion gradient of root exudates to the growing roots. Parkinson, Taylor and Pearson (1963) observed that the effect of root exudates appears to be non-selective. Initially, almost any soil fungus is capable of growing on the root surface and they are just casual colonizers having a limited life on the root-surface because they cannot compete with the true root surface fungi. In most of the cases during the first 2-3 days of the growth of the roots many such casual forms may be isolated, but after 5 days the number decreases when the growth of a limited, typical, surface fungi is observed. Once a stable population of typical root surface forms has been established, it appears to be maintained upto the time of senescence of the root system.

The possible sources of origin of the fungi colonizing plant roots in the early stages of development of the roots, appears to be either the seed coat or the soil. Peterson (1959) and Parkinson and Clark (1964) observed that for crop plants, the seed coat fungi play little part in colonization of the root, the soil fungi are the major source of colonization. Parkinson and Taylor (1961) suggested that colonization of roots by fungi is through the lateral colonization from the soil. However, longitudinal growth down the root from any one point of colonization is restricted in extent.

It is commonly observed that with increasing age of plant roots there is an increasing penetration of fungi into the root tissues. Waid (1957) and Taylor (1962), on the basis of their studies on decomposing rye grass roots and dwarf bean roots respectively, observed little penetration of roots by root surface fungi in the first 40 days of root growth. Subsequently, there was progressive penetration, first of the cortex and then the stele. Among the dominant fungi penetrating deeper in the root tissue *Cylindrocarpon radicicola* and sterile dark forms are worth mentioning (Parkinson, 1965).

Types of organic matter lost from roots: rhizodeposition

Almost all chemical components of the plant have been found to be lost from roots. They can be classified into four different groups depending on their mode of arrival :

(1)   Water soluble exudates - such as sugars, amino acid, organic acid, hormones and vitamins which leak from the root without the expense of metabolic energy.

(2)   Secretions - such as polymeric carbohydrates and enzymes which depend upon metabolic processes for their release.

(3)   Lysate - released when cells autolyse, including cell walls, sloughed off whole cells and with time whole roots.

(4)   Acetylene and CO_2.

Strictly these four classes represent all the types of material lost from the root, but in reality, as microorganisms utilize many of these carbon sources immediately, quantification of losses must also include the consequent microbial biomass and accompanying respiratory CO_2. Thus the term "rhizodeposition" is used to describe all these losses arising from the plant root (Whipps and Lynch, 1985).

The quantities and types of rhizodeposition vary with the age of the plant and position of the root. The initial loss must be from germinating seeds. It is noticed that as a seed imbibes, it immediately loses carbon, often in the form of water soluble exudates and gases and this loss continues as the root and shoot emerge. Carbon loss is related to seed size (Vancura and Hanzlikova, 1972; Naim *et. al.* 1976 and Nelson 1986). Subsequently, as the root grows through the soil, a considerable number of cells are sloughed off from the root cap. Approximately 5000 cells per day are lost from maize (Moore and McClelen, 1983), accompanied by losses of exudates and polysaccharide gel at the rate of 35 g per root per day in sterile cultures (Chaboud, 1983), representing a maximum of 60 mg material lost per gram of root (Newman, 1985). From ^{14}C-labelling experiments, the zone just behind the root tips of both primary and lateral roots appear to be the site of maximum exudation, with the next greatest quanity being lost from the zone of elongation (McDougall and Rovira, 1970). Trofymow *et. al.*, (1987), observed that rhizodeposition of carbon compounds only occurred at root tips.

Most microorganisms in the soil, with the exception of oligotrophs, remain dormant until an input of organic material is made (Wainwright, 1988). The loss of organic material from the roots, as they grow through soil (rhizodeposition), stimulates microbial growth and provides the driving force for rhizosphere development (Whipps and Lynch, 1985, 1986). Thus with time rhizodeposition is measured as root loss and root respiration together with the accompanying microbial biomass utilizing the root derived material, the CO_2 generated by these microorganisms and the portion entering the soil organic matter pool.

The nature of root exudates varies. Carbohydrates, amino acid, vitamins, organic acid, nucleotides, flavonones and enzymes have been identified in root exudates. Saponins, glycosides and hydrocyanic acid have also been reported to be present in exudates. These latter substances, however, have been reported to exert toxic a effect on microrganisms. The exudation of these materials from the root is affected by environmental conditions. Katznelson, Rouatt and Payne (1954, 55) showed that the temporary wilting of plants caused an increased release of amino acids from their roots. Rovira (1959) showed that under condition of high light and temperature, there was increased exudation and this was greatest during the first week of growth.

It appears that the root tip is the zone of maximum exudation (Schroth and Snyder, 1961). Frenzel (1960), however, observed that the exudation of certain amino acids occurs at the root tip whilst that of other amino acids occurs in the root hair zone. This information may be of immense significance to the ecology of microorganisms in the rhizosphere.

Generally, substances present in the exudates are capable of stimulating mycelial growth (Kerr, 1956), stimulating spore germination (Barton, 1957; Schroth and Snyder, 1961; Buxton, 1962) and attracting zoospores of Phytophthora (Bywater and Hickman, 1959). Root exudates of certain plants, on the other hand, contain certain substances which are inhibitory to spore germination and mycelial growth (Buxton, 1962).

Factors affecting microorganisms on root

The following factors influence the colonization of microorganisms on the root:

(a) Soil type
(b) Light regime
(c) Soil moisture
(d) Soil temperature
(e) Cover vegetation

Soil type

The texture of the soil and the hydrogen ion concentration influence the diversity and density of the microorganisms in the rhizosphers. Peterson (1958) noted that *Fusarium spp.* were dominant on the roots from acid soil. *Cylindrocarpon* spp., on the other hand, were abundant on roots collected from alkaline soil. *Trichoderma viride* and *Penicillium* spp. also favour acidic soil. By and large, however, it is a common observation that fungi on the whole are more dominant in acidic soil. Bacteria and actinomycetes along with algae prefer an alkaline pH. There are, however, a number of exceptions and care has to be taken while making such generalizations. Texture wise, well aerated clay with enough organic matter is suitable for a majority of the microorganisms.

Light regime

Light regime, which is important for the growth of the plants, has its own effect on the rhizosphere microorganisms. Harley and Waid (1955 b) noted that in the highest light intensity, *Trichoderma* sp. was dominant on roots whereas *Penicillium* species and *Rhizoctonia* sp. were abundant in low light intensity. They concluded that "The condition of the host plant is a major factor in determining the nature of the surface population of the root system". Rouatt and Katznelson (1960 investigated the effect of light intensity on bacterial flora of wheat root. They observed that of the total number of bacteria, methylene blue reducing forms, glucose fermenting forms and ammonifiers, were much greater on the roots of plants grown under higher light regime. Similarly the amino acid requiring bacteria were more in high light.

Moisture and temperature

Through studies conducted on moisture content, the data indicate that *Penicillium* spp. were dominant under the lowest moisture conditions. *Fusarium oxysporum*, *Cylindrocarpon radicicola* and *Gliocladium* spp. on the other hand were more frequent under the higher moisture regimes (Taylor and Parkinson, 1964). Moderate soil temperature between the range of 15-25°C is normally favourable for most of the microorganisms. Though temperature as such has an effect on the rhizosphere flora, the effect has also to be considered together with the moisture status of the soil environment. The fluctuation in temperature has a direct bearing on the soil moisture content. Both temperature and moisture conditions are inter-linked with each other and hence they have to be taken into consideration jointly.

Cover vegetation

The general condition of the host plant affects the quality and quantity of the rhizosphere microflora. Any change in the physico-chemical conditions of the soil or the environment which affects the host, has its impact on the root microorganisms. As indicated earlier, the substances released from the root in the form of exudates, provide the nutrients to the rhizosphere microorganisms. The factors conducive to the growth of the plants favour rich microflora. The age of the plant is one of the most important attributes in this connection. Normally, the maximum number of rhizosphere microorganisms is reported when the growth of the host is maximum. Sometimes, a second peak is also recorded when the plants enter the senescent stage and the roots start dying. At this stage, the dead root tissue provides an extra energy source for the growth of numerous microorganisms (Mishra and Srivastava, 1973).

To sum up, it is very difficult to study the effect of different factors outlined above on rhizosphere microflora individually. The different parameters operating in soil, rhizosphere and environment are closely inter-linked. A change in one has an impact upon the other. Hence, inspite of so much work being done in the field of rhizosphere, the story is still far from complete and a lot more research is necessary to reach to any meaningful conclusion. The microorganisms in the rhizosphere region are also affected by root respiration. The evolution of CO_2 during respiration of roots alters the pH or the availability of certain inorganic nutrients. In the rhizosphere region, the concentration of CO_2 increases due to the respiration of roots and the microorganisms present there-in. Due to the combined respiratory activities of the two living systems, the root and the microorganisms, the concentration of CO_2 increases drastically. CO_2 thus liberated forms carbonic acid which helps in the solubilization of insoluble, inorganic nutrients not readily available to the plants. This increases the supply of assimiable mineral nutrients and the level of available phosphorous, potassium, magnesium and calcium rises in the region.

In the case of crop plants, CO_2 liberated in the rhizosphere region influences crop nutrition. Soil samples from the root zone produce nitrate more rapidly from humus nitrogen. The total quality of nitrogen mineralized is greater in cropped than in uncropped soil. With the help of rhizosphere flora, the nitrogen availability to plants may be stimulated by N_2-fixing flora and this may be of great importance to agricultural plants. It has been observed that some *Rhizobium* strains are more effective in the presence of root population. In the presence of other bacteria and fungi in the root region, stimulated nodule formation and consequently increased N_2 fixation by *Rhizobium* strains has also been noticed.

An abundance of cellulolytic bacteria in the root zone has been reported. Cellulose-digesting bacteria, particularly Cytophagas and short rods, are frequent in the rhizosphere region. Such bacterial population is responsible for the availability of large quantities of cellulosic tissues which are, in turn, responsible for the degradation of sloughed-off root material.

Root excretions also help in the spore germination of many fungi. Oospores of *Pythium mamillatum* and sclerotia of *Sclerotium cepivorum* are in large amounts in the root region. Water soluble substances liberated from roots activate spore germination.

The effect of rhizosphere microorganisms on plants

As discussed in the preceding pages, a large number of microorganisms differing in their morphological and physiological attributes, form the rhizosphere microflora. The microbial population and activity are much greater in the rhizosphere and in the soil away from the root. It is quite obvious that such a large density of the microbial population in the root region will have its effect both on the development and physiology of the roots. Parkinson (1967) suggested the following four means through which the rhizosphere microorganisms affect the roots and thereby plants in general:

(1)   Effect on root development
(2)   Effect on antibiotic production
(3)   Effect on nutrient availability and
(4)   Effect on root infection fungi.

(1)   **Effect on root development**

The structure, development and morphology of root are affected by the associated microorganisms in the region. Conflicting reports are available on the matter. Both enhanced and reduced growth of roots in the presence of rhizosphere microflora are reported. Pantos (1956) observed an increase in the growth of both tops and roots of wheat plants in the presence of *Agrobacterium radiobacter*. A 65% increase was observed as compared to control plants. Rovira and Bowen (1960), on the other hand, showed a reduction in root hair production and general root growth in clover in the presence of rhizosphere microflora. A decrease in primary root growth, total root growth and secondary roots due to microorganisms was observed in many plants by Bowen and Rovira (1961). It is well known that *Rhizobium* spp. are responsible for inducing curling in legume root hairs prior to infection by the bacterium and subsequently in the forking of roots. It is assumed that growth substances produced by a large number of rhizosphere microorganisms are responsible for such effects. ß indoly lacetic acid is a minor product of many fungi and bacteria common to the rhizosphere. Depending upon the concentration of growth hormones in the root region, stimulatory or inhibitory changes are observed in the root.

(2)   **Effect on antibiotic production**

Besides growth hormones, many microorganisms are capable of synthesizing antibiotics as a metabolic product. For the production of antibiotics a supply of energy rich material is needed by the micro-organisms. In the rhizosphere region, due to root exudates and sloughed-off root tissues, there is no paucity of energy rich materials and the microhabitat is usually very conducive for antibiotic production. This provides a leverage to the antibiotic producing microorganisms for better substrate colonization and to compete over the non-antibiotic producers.

The possibility of antibiotics produced in the root region, being taken up by the roots and transported to the plant body, exists. Once it happens, the antibiotics in the plant body act as systemic inhibitors against many plant pathogens. The deleterious effect of antibiotics on root physiology has also been observed. Norman (1960 a) remarked that even at low concentrations a number of fungal, actinomycete and bacterial antibiotics repress root growth. He further observed that antibiotics injure root cells and cause a leakage of solutes from the irreversibly injured root cells. Polypeptide antibiotics are generally responsible for such activities.

(3) Effect on nutrient availability

The assimilation of nutrients is largely affected by rhizosphere microorganisms. The availability of Mn and PO_4 has been studied in detail. Generally, certain microorganisms induce mineral deficiency diseases due to their metabolic activities in the root region (Brownfield and Skerman, 1950). Timonin (1948) observed that certain bacteria oxidize Mn and render it unavailble to the oat roots and thus results in Mn deficiency which is responsible for grey speck disease in oats. Besides bacteria, certain fungi like species of *Helminthosporium, Curvularia. Cephalosporium, Clado- sporium, Pleospora* etc. are also responsible for Mn oxidation.

The solubilization of phosphate and its uptake is facilitated by some rhizosphere bacteria. Certain fungi are also helpful in phosphate uptake and such fungi form a group in theirself and are responsible for mycorrhizal association. This aspect has been dealt with separately. It appears that organic phosphorous mineralization, using glycerophosphate and nucelic acid as substrates, is more rapid in the rhizosphere.

Solubility of iron is also affected by rhizosphere microorganisms. Iron is utilized in the ferrous form and the conversion of the ferric to the ferrous state is brought about by an increase in acidity or by a fall in the oxidation-reduction potential accompanying microbial metabolism. Sometimes carbohydrate decomposition affects the level of soluble iron. During the process of decomposition, certain organic acids accumulate which chelate the cation.

(4) Effect on root infection fungi

In the soil a large number of microorganisms live and propagate and certain selected ones from such an array of microbial population are drawn to the rhizosphere region. In such a situation, it is quite logical that both saprophytes and parasites have an equal opportunity to colonize the root. The question that arises is who comes first and also the viable effective population of organism(s) is equally important. The population of saprophytic organisms in the soil and the rhizosphere is decidedly higher at a given time than that of parasites. Accordingly, the probability of saprophytes colonizing the root first, always exists. However, in case, potential parasites become established first in the rhizosphere they have every chance to invade and infect the root system. The problem is, however, not so simple. A number of factors act against the infection getting established. The resistance of the root system and the quantum of saprophytic population in the rhizosphere region, inhibit the colonization and establishment of the parasites. In the rhizosphere region, microbial number and activity are enhanced resulting in rapid associative and antagonistic reaction. The rhizosphere saprophytic microorganisms thus form the first line of defense of the plant root system against the attack of plant pathogens.

The prevention of pathogenic infection by rhizosphere microflora has been reported by various workers. *Trichoderma viride* has been reported to act as an antagonist against many root pathogens.

Foliar application of fungicides and insecticides to control the root pathogens, by altering the rhizosphere microbial complex, has also been tried. This may be achieved by changing the metabolism of the plants and encouraging the desired rhizosphere effect or by manipulating the root exudates by the translocation of spranged chemicals. Either way, the rhizosphere micro-environment may suitably be altered to discourage parasitic infection. Vrany, Vancura and Macura (1962) observed that root exudates and rhizosphere microflora could be markedly altered by the foliar spray of antibiotics, growth regulator etc.

Mycorrhiza

Some of the fungi present in soil and in the root region get intimately associated with roots and form a special type of association with the latter. This association is of symbiotic type. Such associations, between fungal hyphae and roots, were known for a long time and a number of publications dealt with the aspect during the latter half of the nineteenth century. However, Frank (1885) for the first time suggested the special name, "Mycorrhiza", for such an intimate association between the hyphae and roots. Frank further remarked that, these organs were always present in the root system of the plants when these were growing in their natural habitats and he also claimed that relatively few roots were free from a fungal sheath.

Earlier, it was thought that mycorrhizal association is the monopoly of a few plant species only and most of the other plants are free from fungal infection. With the lapse of time this concept was drastically revised when a lot of researchers got involved in the study of the phenomenon. All over the world, mycologists, pathologists, physiologists and anatomists got interested in the field and numerous findings on the various above aspects started pouring in. We now know that mycorrhizal association is of common occurrence in almost all the higher plants and the exceptions are a few plant species only. The reason for the involvement of workers of diverse interest is obvious, the morphology and anatomy of the mycorrhizal roots change and the fungi associated with the roots have diverse functions which will be elaborated in detail in the coming pages.

For a number of years, there was a controversy regarding the nature of mycorrhizal association. Many people, out of mere enthusiasm, gave the name mycorrhiza to all the roots where mycelium was associated. No hard and fast

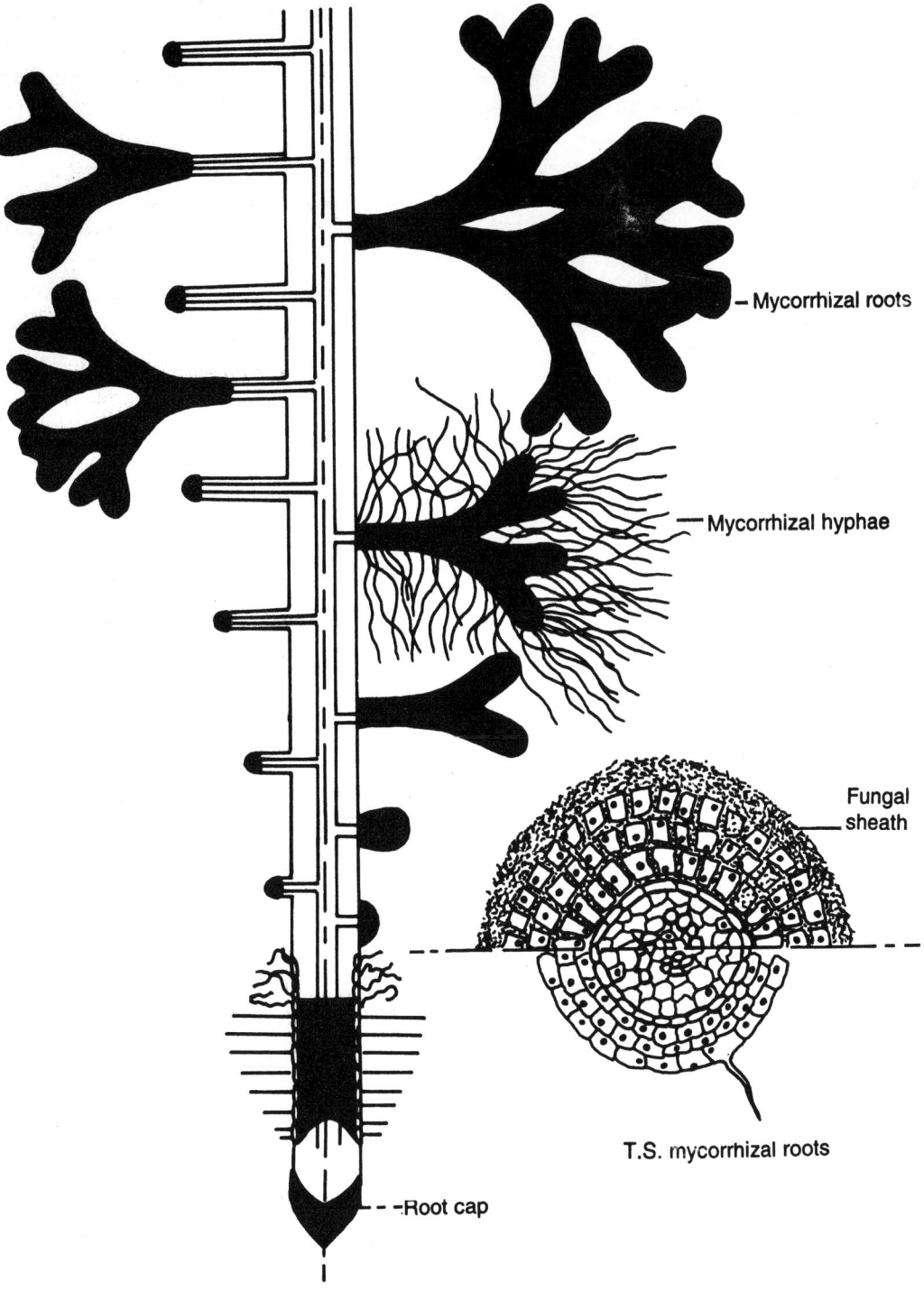

- Mycorrhizal roots

- Mycorrhizal hyphae

Fungal sheath

T.S. mycorrhizal roots

- -Root cap

Fig. 2.1(b) Root region showing mycorrhizal infection

categorization and identification existed. Slowly, however, the picture became clearer and it was subsequently realised that the mycelial association is some-what different from ordinary mycorrhizal association with roots. It was realised that it is not merely the association of the root and the hyphae, but hyphae must enter the root cells and induce anatomical modifications in the root region. As more and more cases were described, the contrast between the forms called mycorrhizas and the mere casual association between fungi and roots became less and less clear. In fact, now it is clear that the term mycorrhiza can only be adopted in such associations of absorbing organs and fungi which are constant in structure and development and are consistently present and functional in natural conditions. With this assumption, mycorrhizal fungi form a separate entity in the system. It is almost universally accepted now that the mycorrhizal condition is a naturally occurring, non-pathological state, in which there exists a prolonged association of symbiotic type. Garrett (1950) remarked "Evolution of the root-inhabiting relationship has culminated in the mycorrhizal association in which a state of true symbiosis has been achieved".

Type of mycorrhizas

Based on the morphology and anatomy of the host plant and also of fungal taxonomy, five broad groupings of mycorrhizas have been suggested. These generally accepted groupings are :

(1) The ectomycorrhizas
(2) the vesicular-arbuscular mycorrhizas
(3) the ericaceous mycorrhizas
(4) the ectendomycorrhizas and
(5) the orchidaceous mycorrhizas

These groupings though are not necessarily definitive categories, but they are useful for examining structure and function. Also all these groupings do share the commonality of a host plant having a fungal associate in intimate, mutualistic symbiosis, where both partners are living and exchange of substances can occur in both directions, i.e. a biotrophic association (Lewis, 1973).

(1) Ectomycorrhizas (E.CM)

The ectomycorrhizas are also called "sheathing" mycorrhizas because of the presence of a distinct shealth or mantle of fungal mycelium that covers the absorbing root. ECM are almost exclusively on woody perennials. The presence of a Hartig net in the cortex is a key diagnostic feature of this association (Marks and Foster, 1973). This net of hyphae extends into the intercelluar spaces of the cortex and they never penetrate the wall into the individual cortical cells. However, sometimes moribund root cap cells may be invaded (Kottke and Obserwinkler, 1986). It is also interesting to note that the fungus never penetrates beyond the endodermis. Fungal mantles vary considerably in thickness-from a few layers of hyphae to

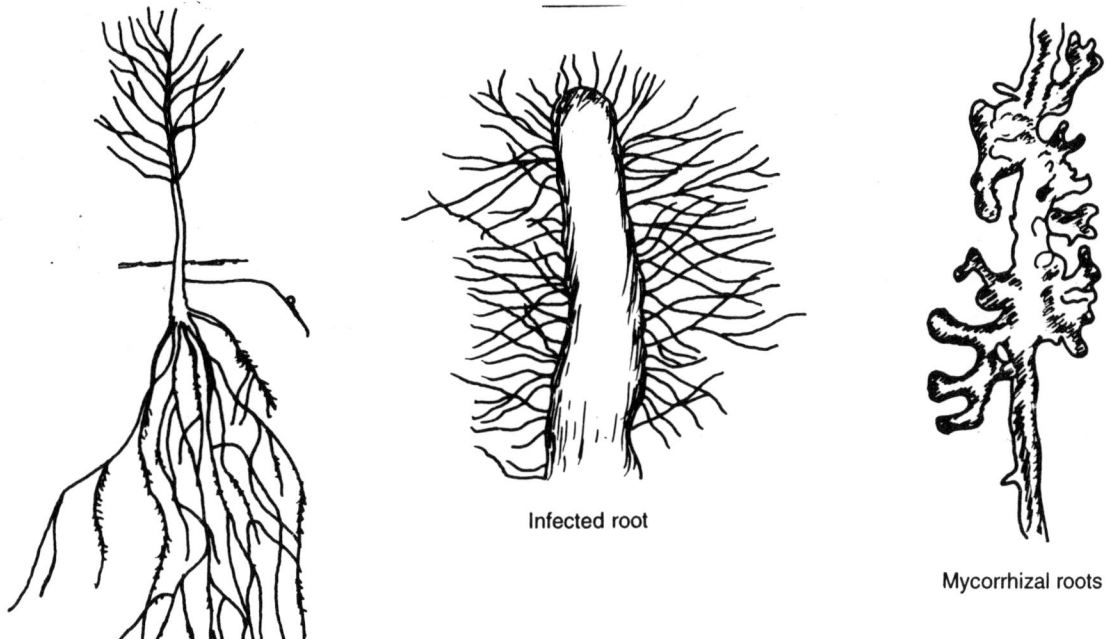

Pinus-mycorrhizal Infected |

Infected root

Mycorrhizal roots

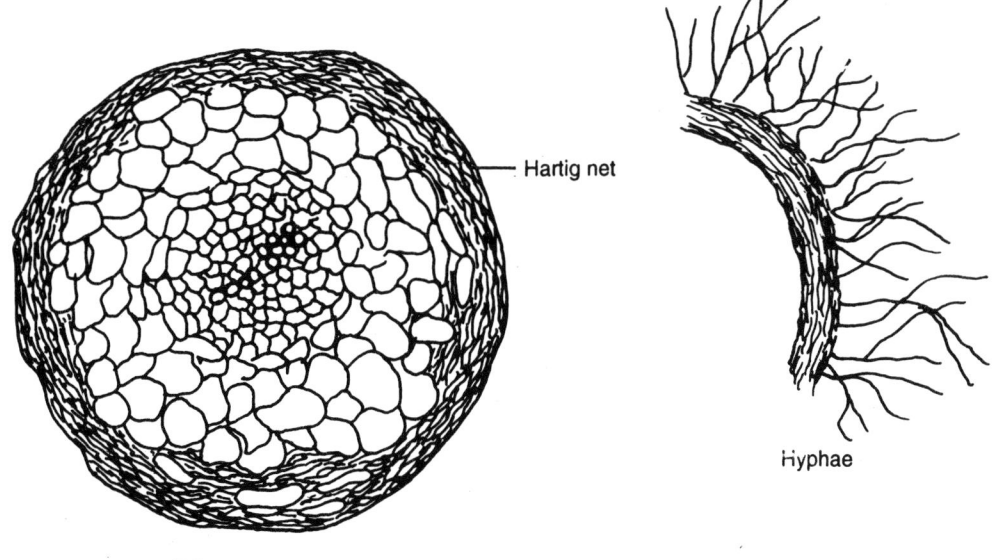

Hartig net

Hyphae

T.S. mycorrhizal root
(Ectomycorrhizal type)

Fig. 2.1(c) T.S. Mycorrhizal Root (Ectomycorrhizal Type)

a radial thickness equal to or much greater than the radial diameter of the root itself. The thickness of the mantle directly affects the surface area of the absorbing organ that is in contact with the soil through change in circumference and may be an indication of the magnitude of carbon utilised in symbiosis. The degree to which hyphae or mycelial strands extend out into the soil from the mantle is another important characteristic that varies considerably. Bowen (1973) and Reid (1984) observed that the presence of mycelial strand and an extensive extra-matrical system may be very advantageous in nutrient uptake in terms of surface area and soil exploitation.

Root growth patterns of the host plant are often altered by ECM development on the root system. It is a common observation that in *Pinus*, proliferation of short roots is stimulated by fungal colonization and dichotomy (bifurcation) of short roots is very prominent in pine ECM. It has also been noticed that the longevity of short roots is increased due to ECM. In natural soils, mainly the shortest roots become modified after fungal infection into mycorrhizal organs. When this happens, the whole of the rootlet is enclosed by a fungal sheath and the apex is surrounded and covered with fungus. Mycorrhizal short roots usually develop many branches whose growth length is restricted, so that the short lateral systems are formed which are completely enclosed in fungus. The appearance of these systems is variable. They are most abundant in the humus layer of the soil where their size in length, diameter and branchiness reaches its maximum.

In transverse section the ECM roots are surrounded by a layer of fungal pseudoparenchyma, the thickness of which varies. Harley (1972) remarked that the thickness of the fungal sheath varies between 20 μ and 40 μ and in axes 300 μ to 500 μ. He further noted that the sheath comprises about 20 to 30 per cent of the total root volume and 34 to 45 percent of the total dry weight. Also, the rootlets have a high content of potassium, nitrogen and phosphate.

The sheath is generally differentiated into an outer layer and an inner layer. The fungal cells with two layers differ. The cells of the outer layer are made up of thickened walls with small or no hyphal spaces, whereas the hyphal structures in the inner layer are loosely packed with wider inter-hyphal spaces and less thickened cellwalls. From the outer layer, a number of hyphal strands extend outside the root and penetrate the soil around the root. These strands, infact, provide hyphal connections with the soil. The inner layer of the shealth is connected to a network of hyphae which run between the cells of the outer cortex. This network of hyphae called the "Hartig net" in most roots extends only between the epidermal cells and the first few layers of the cortex. In certain roots where the diameter is small, hyphae extend into deeper layers, may be, even upto the endodermis.

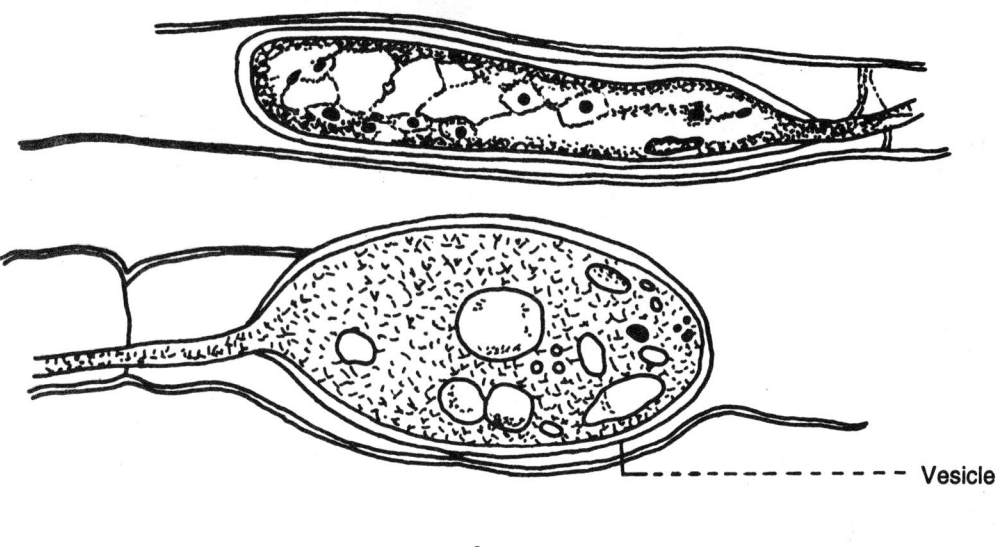

A

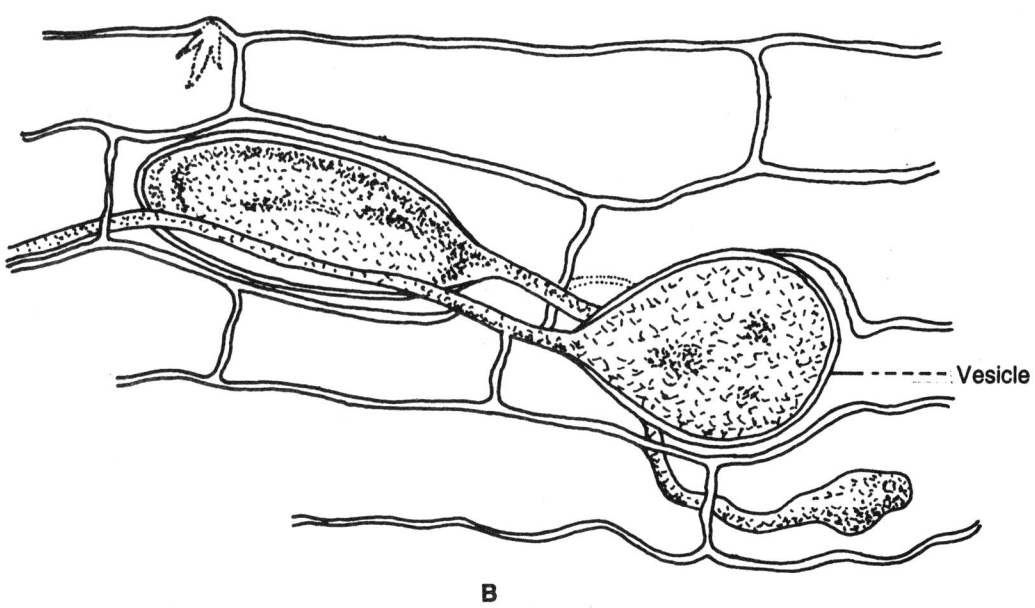

B

Fig. 2.1 (d) Root cells with VAM infection

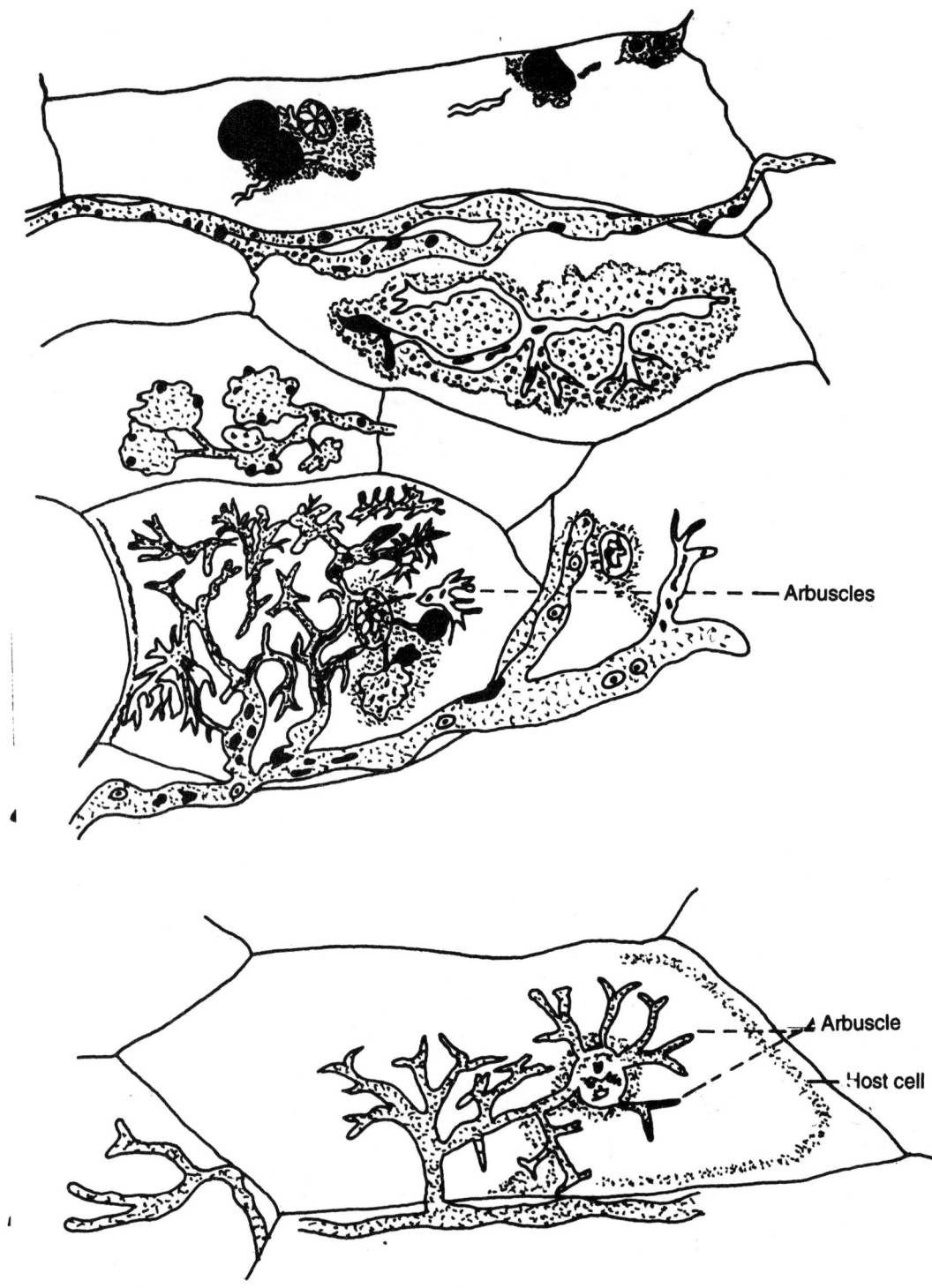

Fig. 2.1(e) VAM associated cells

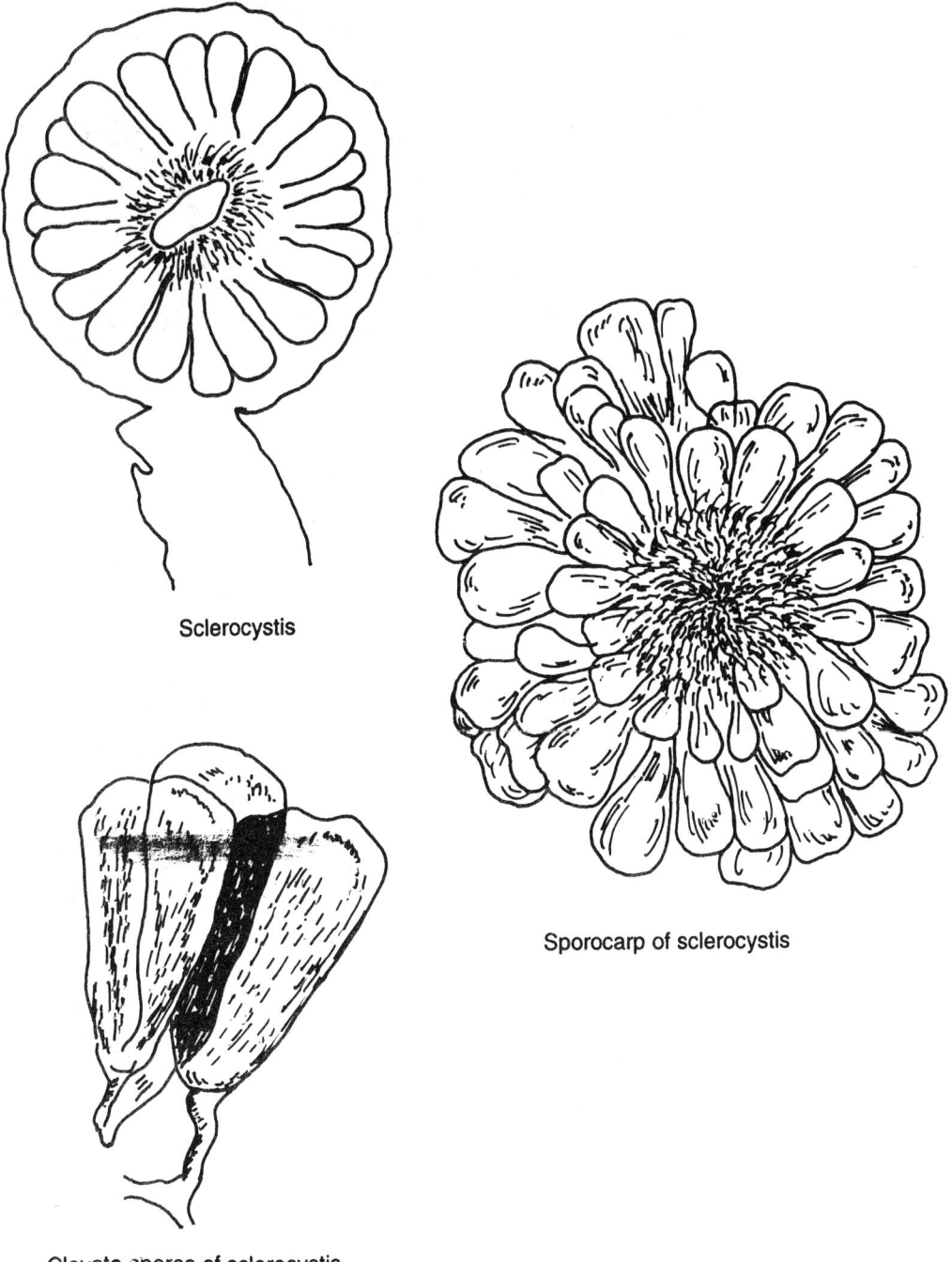

Sclerocystis

Sporocarp of sclerocystis

Clavate spores of sclerocystis

Fig. 2.1(f) Spores of VAM fungi

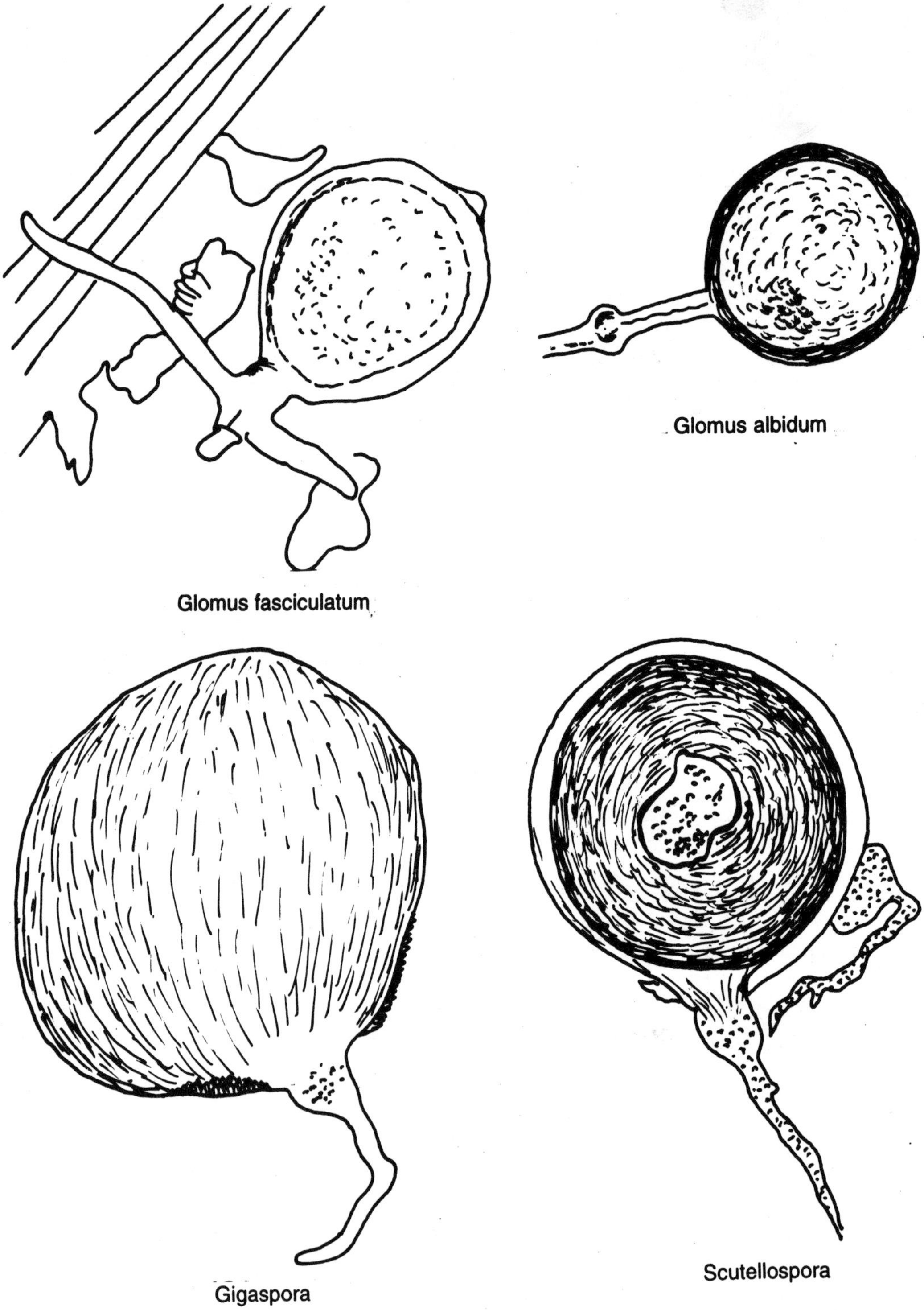

Glomus fasciculatum

Glomus albidum

2.

Gigaspora

Scutellospora

Fig. 2.1(g) Spores of VAM fungi

A very large number of fungal genera have been identified in the ECM association. Most of the fungi forming ECM are of the group basidiomycetes or ascomycetes. In our studies of *Pinus kasiya* about 36 fungal species belonging to basidiomycetes were found to be associated with ECM. The appearance of the fungi was, however, discontinuous. Not all the fungi appeared together, rather a very systemic sequence was noted. The degree of mycorrhizal formation was highest during rainy season and it decreased later on (Sharma and Mishra). It was also observed that short roots of pine were highly mycorrhizal.

Vesicular-arbuscular mycorrhizas (VAM)

Vesicular-arbuscular mycorrhizas are more commonly known as VAM, which has a more appropriate meaning. The term refers to the presence of intracellular structures vesicles and arbuscles - that formed in the root during various phases of development. These mycorrhizas are the most commonly occurring group since they occur on a vast taxonomic range of plants, both herbaceous and woody. In contrast to ECM where a wider range of fungi are associated, in VAM the grouping is deliminated to those mycorrhizas that are formed by the single family endogonaceae of phycomycetes.

In VAM, the infected organs do not always show such marked morphological peculiarities as are seen in other types. Aseptate hyphae, often of large diameter, are observed within the tissues of the host plants. They traverse from cell to cell or between the cells of unspecialized parenchymatous tissue such as the cortex of the root. They donot enter meristems, stelar tissues, chlorophyllous tissues nor other specialised systems of healthy organs. Fungal hyphae within the host cells form coils which may be of one or more turns. In extreme cases, it has been noticed that the coils fill the lumina. It is not necessary that all the host cells contain coils, certain cells may be free from coils. Branches of the main hyphal system which are generally of a smaller diameter penetrate in the cells. Such branches later on produce two typea of structures. They either produce haustoria or complex hyphal branch system which are known as "arbuscles". The branching of the arbuscles is normally dichotomous. The apices of the branches are usually narrow and acute. Sometimes the branches may become swollen at their ends and turn into round fungal bodies - called "sporangioles". The sporangioles may become free within the cells and degenerate inside the cells. The vesicles are generally very large and thick walled. On account of their large size, the cells or the intercellular spaces where they develop, get distorted. Further, as the vasicles develop from multinucleate hyphae, they are multinucleate and heavily loaded with food reserves in the form of fat or oil.

In VAM there is no clear cut sheath formation on the surface of roots. Instead, a loose external weft of hyphae which lies in the soil, is connected with the internal mycelium by connections through the epidermal cells of the host.

Many plants of diverse phyla are colonised by the VAM fungi. From fossil records it is observed that this association existed for millions of years. The earliest geological record is of *Palaeomyces asteroxyli* found in the inner cortex of the rhizomes of *Asteroxylon machiei* from the Devonian rocks of Scotland. The association is also noticed in the decaying stems of *Rhynia* (Kidston and Lang, 1921). Aseptate vesicles bearing hyphae have also been observed in *Mycorrhizinium* (Weiss, 1904), *Lepidodendron* (Seward, 1894) and *Cordaites* (Osborn, 1909). VAM association is continuing in some of the oldest serviving plants like *Psilotum* and *Tmesipteris*. Dowding (1959) remarked that *Endogone* is a common endophyte of living and subfossil roots.

Besides higher plants, many bryophytes and pteridophytes have also been observed to be infected by VAM fungi. Among bryophytes, *Pellia*, *Monoclea* and *Lunularia* have infections of very varied intensity. In *Conocephalum*, *Preissia* and *Marchantia* the fungal infection is localised and the fungus is restricted between the conducting strands and assimilating region. The assimilating tissues are free of fungal infection. In *Anthoceros* the situation is different and the hyphae in this case are intercellular which lateron produce intracelluar arbuscles.

Ophioglossum spp., *Botrychium ternatum* and *Lycopodium spp.* are mycotrophic genera in pteridophytes. In *Dryopteris* and *Phyllitis scolopendrium* also the fungal infection has been noticed.

In conifers, endotrophic mycorrhiza involving aseptate phycomycetes mycelia is very common. In Pinaceae, however, generally ectotrophic mycorrhizas are prevalent. In genera, like *Cryptomeria*, *Sequoia*, Taxodium, *Cunninghamia*, *Juniperus*, *Agathis*, *Aurocaria*, *Ginkgo*, *Cephalotaxus*, *Torreyas*, *Cupressus*, *Biota*, *Thuja* and *Thujopsis* etc. VAM type of mycorrhiza is widely encountered.

In angiospermous families, endotrophic fungal infection is wide spread. Boullard (1956) stated that families like juncaceae, cyperaceae, cruciferae are devoid of mycorrhizal assoication. Harley (1972), however, predicted that even in these families some species or genera may be found to form mycorrhizas. Lateron, this prediction was found to be correct and mycorrhizal association has now been reported in certain plants of cyperaceae and cruciferae (Mishra *et. al.*, 1980).

3. **Ericaceous mycorrhizas**

Most of the plants of Ericales possess mycorrhiza. With some minor finer variations, the mycorrhizal infection found in the plants of the group generally conform to one type. In all cases, mycelial penetration into cells and tissues by septate hyphae is noticed. There also is found a considerable development of an external mycelium on the root surface in the adjacent soil.

Ericaceous mycorrhizas are further divided into two groups : ericoid and arbutoid types.

Ericoid mycorrhizas

The characteristic feature of the roots of many Ericaceae is their fineness. The ultimate branches of the roots consist of narrow central stele and few, sometimes form only a row of coritical cells. The rootlets are surrounded by a loose weft of septate mycelium. The mycelium grows and penetrates inside the cortex and enters the cortical cells in many places. There exists some difference in the diameter of the mycelium present inside and outside the host cells; generally those intracellular in position have a larger diameter. The infection is confined to cortical cells of the ultimate branches and the oldest roots are either free from the infection or are less infected.

In many respects the development of ericoid mycorrhizas is similar to the mycorrhizas in forest trees.

Ternetz (1907) and Rayner and Livisohn (1940) isolated *Phoma radicis* from the infected roots of ericaceous plants and they assumed that this fungus formed the mycorrhiza.

Arbutoid mycorrhizas

The typical representative of the group *Arbutus* is provided with two types of roots i.e. long and short branched ones. The short branches are typically mycorrhizal. The mycorrhizas are more like those of ectotrophic organs of forest trees. The short roots are endogenous, produced from the long roots and they get infected as they pass through the cortex. The short roots are covered with a mucilaginous coat and it is this which is first colonized by fungal hyphae. The layer of hyphal thickness is several cells in depth, the rootlets swell and a tubercle covered by a dense pseudoparenchyma is formed. The cortex of the tubercle is soon penetrated by the fungus and coils are formed in the cells. After sometime, the cells become clumped together, get digested and subsequently disintegrate.

In *Monotropa*, the roots are covered with fungal sheath which is greatest in humic soils. Hyphal branches from the sheath penetrate the tissues to form a network similar to a Hartig net, in the cellwalls of the outer cortex. At later stages, these hyphae and the sheath hyphae form haustoria which penetrate into the lumina of the outer most cells of the host. Subsequently, the free tips of the haustoria in the region of the host nucleus get digested and the disorganised haustoria become completely sealed off by the host. Franke (1934) suggested a species of *Boletus* to be responsible for forming mycorrhiza in *Monotropa*.

4. Ectendomycorrhizas

These mycorrhizas form a Hartig net in the cortex of the root but develop little or no sheath. Intracellular penetration of cortical cells also takes place and in this respect they are similar to the arbutoid mycorrhiza. Ectendomycorrhizas are commonly found in Pinaceae but the association is limited to forest nurseries. The fungi associated with such types of mycorrhizas are called E-strain and these are most likely the imperfect stage of ascomycetes. The peculiar feature of such fungi is that they form ectendomycorrhiza in some tree species and ECM in other tree species. In many cases the ectendomycorrhizas formed initially in nursery seedlings are later on replaced by ECM.

5.   **Orchidaceous mycorrhizas**

The fungal association is of endomycorrhizal type. The fungus penetrates the cell-wall and invaginates the plasmalemma forming hyphal coils within the cell. Once the plant is invaded, the spread of the fungus may occur internally from cell to cell. The internal hyphae eventually collapse or are digested by the host cell. Harley and Smith (1983) remarked that since the symbiosis forms an external network of hyphae, it is possible that the fungal hyphae function in nutrient uptake as in other mycorrhizas. In this way, the coarse root system of orchid would be supplemented by the increased absorbing surface area of the hyphae.

The orchid mycorrhizas are somewhat different and unusual in the sense that the roles are reversed in this symbiosis. In the early stage of the orchids life cycle, this is more pronounced when the plant is not autotrophic.

All orchidaceae, at one stage or other of their life cycle are infected with fungi under natural conditions so that their absorbing organs maintain fungal mycelia within their cortical cells. The orchids as a group have in common a kind of mycorrhiza of their own and their fungi, too, in many ways constitute a definable group. They are either basidiomycetes, often with clamp-bearing hyphae or imperfect fungi of the genus *Rhizoctonia*.

The seeds of orchids are very minute in size and they are produced in large numbers in each capsule. The embryos typically lack cotyledons. They are enclosed at maturity in a loose, thickened testa which may vary greatly in form and ornamentation. The small size of embryo and the absence of food reserves in the seed, demand that the greater part of the food material utilized, upto the period when photosynthesis begins, must come from the substrate. In the early stages of germination, the plantlets get infected by fungal hyphae. It is interesting to note that the entry of fungal hyphae into the young seedling at germination is necessary for the initiation of further growth in orchids. Normally the young protocorm is first to be infected, however, the young roots are not directly infected. The hyphae of the protocorm and the infection of roots takes place anew from the soil.

The outstanding properties of the fungi of orchid mycorrhiza which differ from those of ectotrophic mycorrhiza are their ability to utilize complex sources of carbon, their rapidity of growth and their ability to compete as saprophytes with other organisms. Also, the orchid mycorrhizas play an active and essential part in the nutrition of orchids in both saprophytic and autotrophic phases of the life cycles of the hosts.

Function of mycorrhizas

Mycorrhizas play an important role in the soil. In many cases the mycorrhizal system actually bridges across the rhizosphere and provides an organic link between the root and the bulk of soil. In addition to their enhanced capacity in acquiring nutrients from the soil, which is now an established fact, they are also helpful in many other ways as described below:

(1)   Nutrient acquisition
(2)   Water use
(3)   Root protection
(4)   Interactions with other microorganisms
(5)   Impact on carbon cycle

1.   Nutrient Acquisition

The nutrient absorbing system in mycorrhiza is drastically affected. The effective surface area for ion uptake increases many fold. The movement of inorganic ions to absorbing surfaces in the soil, both in case of root surfaces and hyphal surfaces, occurs by diffusion and mass-flow. Nye (1979) suggested that the rate of uptake by the root will depend on the demand by the absorbing organ, the size and distribution of the absorbing system and the mobility of the nutrient in soil. Positive growth response to mycorrhizal development can generally be expected where the concentration of some nutrients is extremely low in the aqueous phase, but some solid or unavailable form exists in reserve (Tinker, 1975). Clarkson (1985) observed that the extension of hyphae from the mycorrhiza, beyond the depletion zone of the root, effectively allows the absorbing system to access the bulk soil solution concentration.

As the increase in surface area of the mycorrhizal roots is supposed to be primarily responsible for enhanced nutrient uptake, many factors like physical extent of the hyphal system, length of time for which it is functional, soil conditions under which it is functional, absorbing power of the hyphae and exploitation of nutrient rich sites in the soil, influence their activity.

Based on the researches of many workers it has almost been established that the surface area of mycorrhizal roots in enhanced. Abbott and Robinson (1985), reported over 3000 cm. of hyphae per cm. of infected root of subterranean clover at 5 weeks, when inoculated with *Glomus fasciculatus*. It has also been observed that almost 96% of hyphae are active (Sylvia, 1988). The maximum length of hyphae per unit length of root however, varies for different stages of the development of the root, and also with respect to the symbiont associated. Abbott and Robinson (1985) further observed that peaks in maximum length of hyphae per unit root length are between 4 and 5 weeks, while Sylvia (1988) showed a peak and levelling off of active hyphae after 8 weeks, as root colonization continued to increase upto the last date of harvest at 13 weeks. Bethlenfalvay *et. al.* (1982 b) observed a similar increase and peaking of external hyphae at 10 weeks, whereas internal fungal biomass continued to increase till 19 weeks. VAM has been shown to transport substances upto a distance of 8 cm. from the root (Pearson and Tinker 1975; Rhodes and Gerdemann, 1978), and the depletion zone around roots has been shown to be increased by 1-2 mm by mycorrhizal infection (OWUSU - Bennoch and Wild, 1979).

The development of the hyphae and mycelial system connected to the sheath of the ECM is variable and suggests that the characteristics of this external system should be reflected in the relative effectiveness of the mycorrhiza in nutrient uptake (Harley and Smith, 1983). In sheath forming mycorrhizas the thickness of the sheath affects various features. The surface area increases with the circumference of the root, there is possible exclusion of other microflora or microfauna and the biomass utilizing carbon and the capacity for storage of nutrient elements both depend on sheath thickness. The sheath in ECM is typically 20-40 μ thick and may comprise 20-30% of the volume of the rootlet (Harley and Smith, 1983). Normally, the sheath is responsible for doubling of the surface area in contact with the soil.

The expected life of hyphae and mycelial strands in soil and the length of time for which they are functional in the root, vary. In VAM, various estimates indicate that an infection unit remains functional in the root from about 4 days (Cox and Thinker, 1976) to 15 days (Brown *et. al.* 1975). The functional period is based on the time that the arbuscles remain intact in the cortical cell - the internal phase that seems most appropriate for nutrient exchanges between host and fungus (Bonfante-Fasolo, 1984). The condition is somewhat different when we consider the root as a whole because new cortical cells are continuously infected, thereby retaining an effective nutrient exchange interface between the root cortex and the extra-matrical hyphal system. In case of ericoid mycorrhizas, Read (1983) observed that the active life of the infection unit is probably not more than 5 to 6 weeks.

By and large, the processes of absorption of nutrients by mycorrhizal fungi are similar to those of roots of higher plants. Also, the absorbing capacity of mycorrhizal roots is much greater. Harley and McCready (1950) noted that the phosphate absorption rate by ECM was five times greater than the non- mycorrhizal roots. It is also observed that biochemical mineralization and chelation by the fungal components of mycorrhizas could be important in solubilizing certain nutrients in the rhizospheres. Bartlett and Lewis (1973) noted that the surface acid phosphatases present in ECM and VAM are useful in utilizing complex phosphates. Higher phosphatase activity in the germ tube and hyphae of VAM fungus *Glomus mosseae* has also been reported (Dodd *et. al.* 1987).

In the root region, the situation is, however, very complex. In addition to increase phosphatase activity due to mycorrhizal fungi, plant roots and other soil microorganisms also produce acid phosphatases. In such situations, it is very difficult to assess the role of the different components of the root region in the production and activity of acid phosphatses. Also, the exact contribution of mycorrhizas in enhancing phosphorous availability is difficult to determine. In addition to mycorrhizal fungi, other microorganisms present in the rhizosphere region also have the ability to modify the availability of phosphorous by producing metabolites which release phosphorus from insoluble phosphates (Ca, Al, Fe), by the formation of soluble complexes with metal ions (Woldendorp, 1978).

Besides phosphorus, the role of mycorrhizal fungi in utilizing Al and Fe salts of inositol hexaphosphate in pure culture has also been suggested (Mitchell and Read, 1981). The prevalence of the calcium salts of oxalic acid on the outer surface of the hyphae of forest litter is another aspect which is ascribed to mycorrhizal fungi (Graustein *et. al.*, 1977).

It is now becoming clear that many nutrients can be taken up by mycorrhizal hyphae and transported to the root. Much work has been done, specially with respect to phosphorus uptake, translocation and storage and to a lesser extent with nitrogen. Nitrogen, specially ammonium has received increasing attention. It has been demonstrated that mycorrhizal fungi can enhance the nitrogen nutrition of plants (Smith *et. al.*, 1985; Barea *et. al.*, 1987).

Studies have also demonstrated that VAM fungi help in the uptake and transport of P, S, Zn, Cu, Ca, N, K and Sr. ECM is helpful in P, N, K, Ca, Cl and Zn (Harley and Smith, 1983). Alexander *et. al.*, (1984) observed that orchidaceous mycorrhizal fungi are useful for P and Read (1983) noted the usefulness of ericaceous mycorrhizal fungi for P and N.

It seems that cytoplasmic streaming coupled with mass-flow are the mechanism of transport in fungal hyphae (Tinker and Gildon, 1983).

2. **Water Use**

Hypothetically, the increased surface area provided by the extra- matrical hyphal system, can provide a direct pathway for water uptake to the root. In soil - plant - atmosphere continuum, the flow of water is driven by a potential gradient established by transpiring leaf surfaces. This, however, may get modified by a series of resistances and capacitances in the continuum. Water moves in the soil by bulk or mass-flow and hence the change in water content has an impact on conductivity.

The importance of hyphae in the overall water economy of the plant is a controversial issue. Sanders and Tinker (1973) and Graham and Syvertsen (1984) suggested that hyphal transport alone is probably responsible for an increase in water uptake by mycorrhizal plants. Allen (1982) and Hardie (1985), on the other hand, believe that additional surface area provided by the extra- matrical hyphae does offer a reasonable explanation for increased water uptake by mycorrhizal plants. In case of ECM, the segregation of hyphae into strands may account for greater fluxes of water to the root.

Whatever the controversies might be regarding the ability of hyphae and mycelial strands to conduct water at low soil moisture or in very coarse-textured soils, it has been convincingly demonstrated by a number of workers that the mycorrhizal plants are better adapted to survive in water scarce soils. Read (1977) suggested that the formation of mycorrhizas by *Cenococcum* with *Helianthemum* was an adaptation to dry soil conditions. Cromer (1935) and Harley (1940) also suggested that ECM were more resistant to desiccation than non-mycorrhizal roots.

3.   **Root protection**

A close association of fungal hyphae with the plant root changes the structure and morphology of the root. It is clear now that the root environment is very complex, where soil and *rhizosphere microorganisms* and *mycorrhizal fungi* live and interact with each other. Interactions between mycorrhizal fungi and other microorganisms in the micro-environment is of a very complex type. In this process the potentially deleterious organisms in the region are also affected.

Marx, as early as 1969, demonstrated that ECM fungi could provide protection to roots from the pathogen *Phytophthora cinnamomi*. In his opinion, both antibiotic production by the fungi and the physical barrier provided by the sheath, were mainly responsible for protection of the root.

The mycorrhizal fungi are supposed to influence root pathogens in the following ways :

1. Competition for the uptake of essential nutrients in the rhizosphere and the root surface.
2. Competition for actual site of interactions in and on the root.
3. Alteration of physiology of the host plant.
4. Presence of a physical barrier to infection.
5. Selection of favourable microflora by changes in root exudate products.
6. Production of toxic or inhibitory substances.
7. Nutrient absorption system compensation for damage to root by disease (Harley and Smith, 1983; Marx, 1975).

Thus it appears that, not only are the microbial interactions in the root region complex, but the mechanisms influencing the root pathogens by mycorrhizal fungi, are also equally complex and are of varied nature. It is equally possible that not only one, but a combination of the above mechanisms may operate simultaneously to influence the root pathogen. In such cases, it is very difficult to pin point any mechanism effectively responsible for the check of the pathogen and on the practical sides, even to suggest a definite mechanism for use in the control of the pathogen. The whole story is really very complex.

Inspite of all these deficiencies in understanding the mechanism, reports are available on the beneficial role mycorrhizal fungi play with respect to root pathogen control. Enhanced nutrient status of the plant for VAM Citrus (Davis, 1980, Davis and Menge, 1980), high arginine levels in VAM tobacco that were inhibitory to pathogen chlamydospores (Baltruschat and Schonbeck, 1975) and Cell wall thickening in VAM onion (Becker as cited by Bagyaraj, 1984), have been suggested to be means for the control. Recently, Chakravarty and Unestam (1987) showed the complete inhibition of *Cylindrocarpon destructas* present in the rhizosphere by certain ECM fungi of *Pinus sylvestris*. *Laccaria laccata* and *Pisolithus tinctorius* were observed to be the most effective. In their opinion, the utiliztion of root produced substrate by the mycorrhizal fungi, may have limited the growth of pathogens.

4.   **Interaction between mycorrhizas and other microorganisms**

Relatively few studies are available on mycorrhiza-microbes interactions. Both, mineral nutrition and carbon allocation of host plants, alter the chemical and physical environment which undoubtedly has its effect on root associated microbes in the soil. Bagyaraj (1984) observed that in most of the studies on tripartite association of leguminous plant, VAM fungus and *Rhizobium*, mycorrhizas generally improve nodulation and nitrogen fixation. Smith *et. al.* (1985) noted that mycorrhizal fungus encourages the production of glutamine synthetase in VAM root, which accounts for the increased ammonium uptake from soil. Pacovsky *et. al.* observed an improvement in the shoot and root dry weight of sorghum when VAM fungus and *Azospirillum* were inoculated . Dual inoculation with both organisms had an additive effect on shoot dry weight. Phosphate-solubilizing bacteria in the rhizosphere of mycorrhizal plants have a special significance. Reports are available showing that such bacteria interact with VAM for a better exploitation of a poorly - soluble phosphorus source. Better growth, increased dry matter and enhanced uptake of phosphorus from the soil were observed when VAM fungi and phosphate solubilizing bacteria were inoculated together (Raj *et. al.*, 1981; Piccini and Azcon, 1987). Inhibitory ability of *Trichoderma* against fungal colonization is altered in ECM infected roots (Summerbell, 1987). It has also been observed that bacteria and actinomycetes associated with ECM fungi, affected the development of mycorrhizas through cellulolytic and pectolytic enzymes (Dalim *et. al.*, 1987).

5.   **Mycorrhizas and carbon balance**

Schulze (1982) discussed in detail the importance of carbon allocation between shoots and roots, for survival and competitive ability. Undoubtedly, the development of mycorrhizas affects the carbon balance of the plant. Vogt *et. al.,* (1982) calculated the mycorrhizal fungal component in silver fir forest and observed that approximately 14-15% of the net primary is comprised of fungal components, which roughly represented 2800-4100 kg ha^{-1} of dry matter. Working on a similar line, Fogel and Hunt (1979) noted that the annual fungal throughput in a second-growth Douglas fir stand was about 50% of the total annual stand throughput of 30300 kg Ha^{-1}

Some of the above informations, however, are not very accurate since the above estimates do not take into consideration the cost of respiration of the fungal system and also may not include all of the turnover of fungal structures. Much documented experimental evidence is still needed to reach a meaningful conclusion.

Chapter 3

Soil microorganisms and organic matter decomposition

From different sources organic matter is added to soil. Vast amounts of above ground tissues and underground portions of the plant are mechanically incorporated into the soil. It has been observed that in most natural vegetation, the amount of organic matter in the soil system remains approximately constant from one year to another, despite large seasonal additions from falling leaves and other parts of the plant. Besides plant remains, animal tissues and excretory products also contribute to the bulk of organic matter. The soil is also rich in microbial population of various types. The microorganisms die and the dead cells subsequently serve as a source of carbon for succeeding generations of microbial population. The type of substances thus continuously being added and incorporated into the soil are of diverse nature. The chemical composition of organic detritus varies from very simple to extremely complex. The physical and chemical nature is extremely variable and heterogenous. All such substances sooner or later depending upon their structural heterogeneity are subjected to microbial attack and they form the food of the soil microorganisms.

According to Alexander (1961) the organic constituents of plants are divided into 6 broad categories :

(a) the most abundant chemical constituent - cellulose varies from 15 to 60% of the dry weight;

(b) hemicellulose - commonly making up 10 to 30% of the weight;

(c) lignin - normally contributes 5 to 30% of the plant,

(d) the water - soluble fraction - simple sugars, amino acids and aliphatic acids are included in this category and they contribute 5 to 30% of the tissue weight;

(e) ether and alcohol soluble constituents - fats, oils, waxes, resins and a number of pigments are the major constituents of this category;

(f) proteins which are generally rich in plant nitrogen and sulphur.

The mineral constitutents vary from 1 to 13% of the total tissue.

In general, litter production in a stable forest is often constant. The amount of leaves accumulated annually on the floor of a given system is usually not much different in different years. Minor fluctuations one can ofcourse expect. On the other hand, the fall of branches or stems is very erratic. Bray and Gorham (1966) found for four successive years the values of leaves etc. to be 2.8, 3.2, 3.2 and 3.1 metric tons/ha, comparable figures for stems, including bark, were 0.6, 3.2, 0.5 and 0.8 metric tons/ha. The variations are also quite significant for different vegetational types and their location. In alpine and arctic forests the amount is usually very low where as in mature secondary forests and in mixed forests of the tropics the quantity is significantly very high. Climatic changes have also great impact on litter accumulation. Through out the year some amount of litter of varying nature falls on the surface of soil. Starting from bud scales to flowering parts in the life cycle of a plant, the dead parts are detached and they are deposited in the form of litter. Some times, during the year the amount of litter may be, though, considerably less but it is associated with an increased activity of the litter and soil organisms.

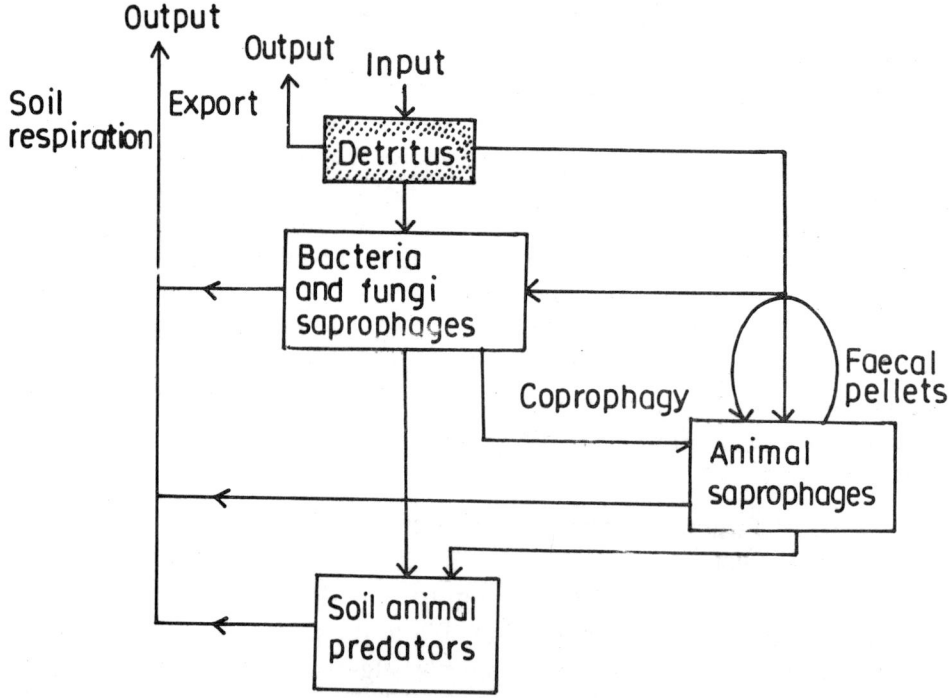

Fig. 3.1(a) An integrated model of the functioning of the organisms of the decomposition system

In plant tissue, the bulk of material is composed of cellulose, hemicellulose and lignin. In woody tissues the major constituents are cellulose, lignin and to some extent hemicellulose and water soluble materials are in small quantities. With the ageing of the plant the water soluble constituents, proteins and minerals decrease and the amount of cellulose, hemicellulose and lignin increases. Thus the litter, in general, is composed of mixed and highly diverse substrates which subsequently form the substrate for decomposition and mineralization of carbon by soil microorganisms.

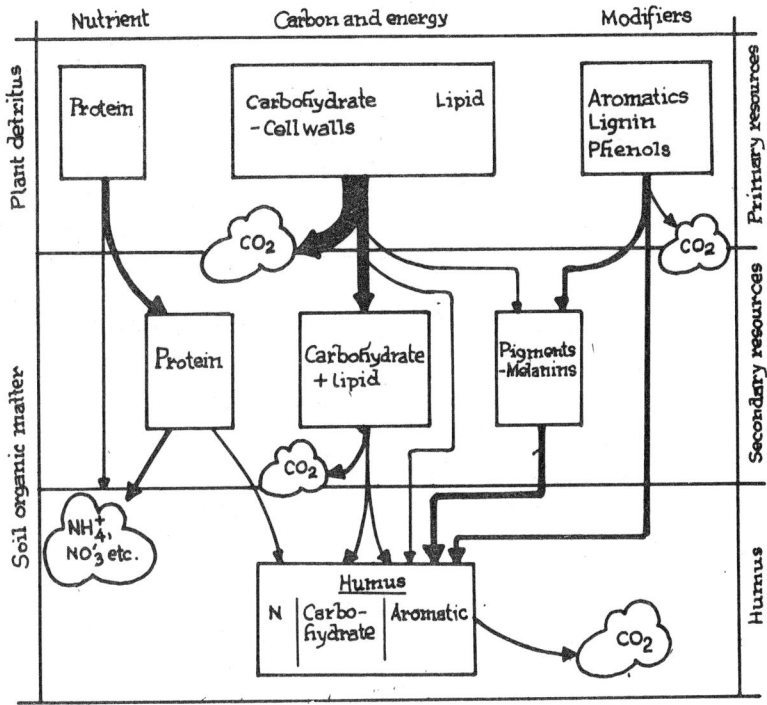

Fig. 3.1(b) Major Chemical pathways of decomposition

The amount of underground material added annually to soil is normally difficult to estimate. Bray and Gorham (1966) indicated that the net production of below ground parts is usually about two-third the weight of leaves produced in temperate forests. The amount is a little high in tropical forests where the contribution of under ground parts is nearer to one-third of the leaf production.

Invasion of plant tissue by microorganisms

It is interesting to note that the colonization of plant tissue by microorganisms takes place much before they are shed. Every part of the plant is reasonably well colonized by a number of microorganisms when it is still attached to the plant. Much work has been done on the colonization of leaves when they are still green and attached to the plant. Phyllosphere has attracted the attention of a number of workers. Like the rhizosphere of root, the surface of the leaf known as phyllosphere is also rich in nutrition. The metabolites are released from the leaf surface in the form of leaf exudates which contain a variety of nutrients. The substances available in the exudates are almost the same which are present in the rhizosphere and it has already been described in the earlier chapter on

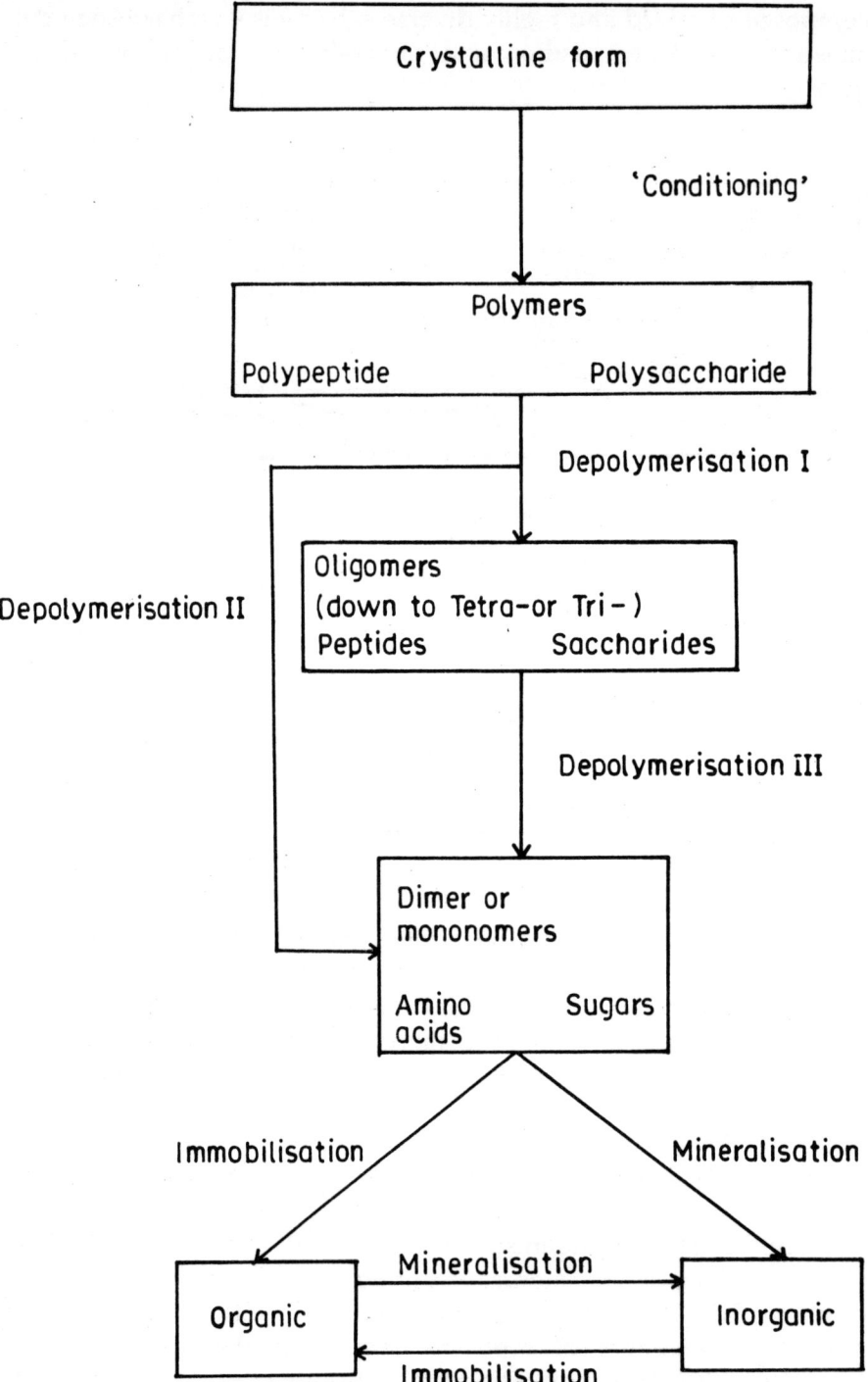

Fig. 3.1(c) Model of the main steps in the catabolism of polymeric substrates by decomposer organisms

rhizosphere. However, it will not be out of place to make a mention of the nutrients present in exudates. Generally varying amount of sugars, carbohydrates, amino acids, phenols, proteins, hormones etc. have been found to be present in the exudates (Mishra and Srivastava, 1970). The nutrients thus provide enough opportunity for a number of microorganisms to colonize the leaf surface. The type and quality of phyllosphere microorganisms vary at different stages of the development of the plant (Mishra and Strivastava, 1970). The net result, however, is that by the time the leaves become senescent and fall from the tree they are already loaded with a number of microorganisms. In the true sense, the break down of the leaf tissue begins much ahead to what is actually known as litter decomposition. The phyllosphere microflora is of a diverse nature ranging from simple saprophytes to weak parasites and finally true parasites. Certain leaf surface flora like *Pullularia* and *Cladosporium* have been noticed to decompose simple hexoses and pentoses. The leaf surface microflora is relatively less in cold temperate environments, however, in moist tropical plants the population is considerably high (Mishra and Srivastava, 1971). In our studies on different plants we have observed that a large number of fungi colonize the leaf surface and the population structure of the microbes consists of a variety of forms. The nature of microorganisms varies from plant to plant and also for the same plant at different stages of development (Mishra, 1967).

Many fungi associated with the leaf surface actively participate in the decomposition of the substrate and they become more active once the plant material is detached and forms litter. Webster (1956, 57) noted that the moribund tissue of *Dactylis* is invaded first by the primary saprophytes, *Cladosporium*, *Epicoccum*, *Alternaria*, *Leptosphaeria* and *Pleospora* which later on advance up the stem as the new leaves unfold. He also observed that different saprophytic fungi are associated with different nodes. Working with *Pteridium* petioles Frankland (1966) observed that under natural conditions, invasion and considerable decomposition would seem to occur before the petioles become incorporated with the litter proper. In tropical rain forests considerable disintegration of branch systems normally is achieved before they finally fall to the ground. This disintegration is primarily accomplished by fungi and white ants.

Kendrick and Burges (1962) during their study on the decomposition of pine litter observed that in addition to numerous saprophytes, a few parasites like *Conosporium*, *Lophodermium* and *Fusicoccum* are also associated with the needles while they are still alive. The activity of the parasite, however, differs. No activity of *Conosporium* is observed once the needles are shed, however, the situation is different in case of *Lophodermium* and *Fusicoccum*. *Lophodermium* remains active upto 6 months of the fall of the needles and *Fusicoccum* which is relatively less active initially becomes more vigorous after the needle dies and falls and the fungus sporulates heavily 3 to 5 months later.

The major constituents of plant litter as indicated earlier are cellulose, hemicellulose and lignin. Besides these three, other polysaccharides which occur in varying amounts are starch, the pectic substances, inulin, gums and chitin. It will be

(A)

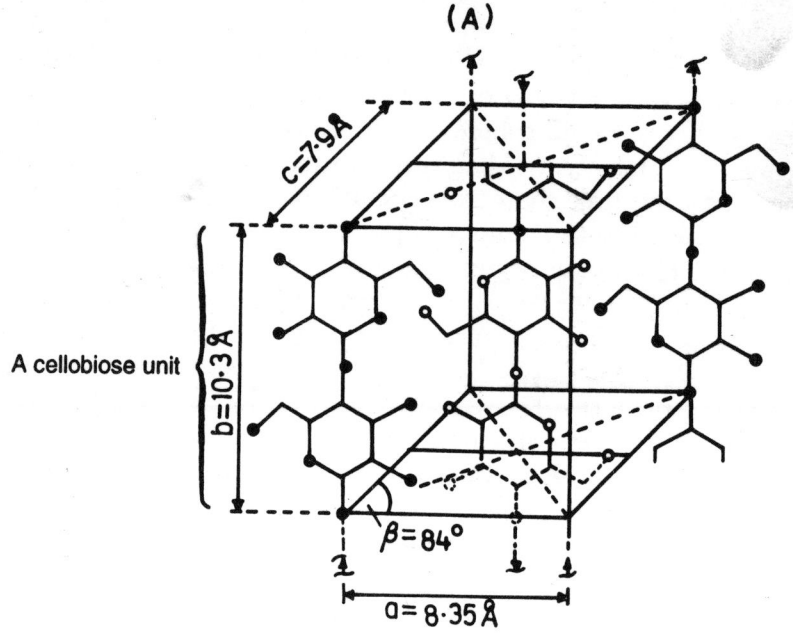

A cellobiose unit

(B)

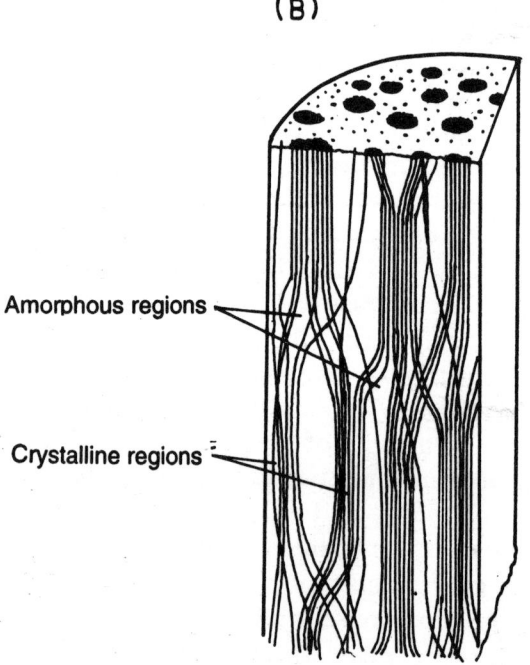

Amorphous regions

Crystalline regions

Fig. 3.1(d) Mycrofibrillar structure of cellulose (A) Hydrogen bonds between adjacen t chains (B) Arrangement of elementary fibrills (L.S.)

appropriate to discuss in short the decomposition of the above constituents.

Cellulose

Microbiology of cellulose has a special significance because a large part of the vegetation added to the soil in the form of litter is composed of cellulose. Structurally cellulose is a carbohydrate composed of glucose units. The units are bound together in a linear sequence forming a long chain. Glucose units are linked by ß - linkages at carbon atoms 1 and 4 of the sugar molecule. In each molecule the number of glucose residues varies between 1,400 and 10,000. The number of sugar units per chain and the molecular weight of cellulose, however, differs for different plant species. The range of molecular weight is normally noted to be between 2,00,000 to almost 2 million.

Cellulose is of common occurrence in almost all the plant species. It is restricted to the cell wall where it is present in the form of submicroscopic rod shaped units known as micelles. The latter in turn are arranged into a large structure, the microfibril. About 10 to 20 micelles are seen to occur in a microfibril. Other polysaccharides like xylans, mannans and polyuronides are also associated with the cellulose of the plant cell wall. Arabans and galactans are sometimes found in small quantities. Such polysaccharides that are structurally linked with the cellulose of the cell wall are known as cellulosans.

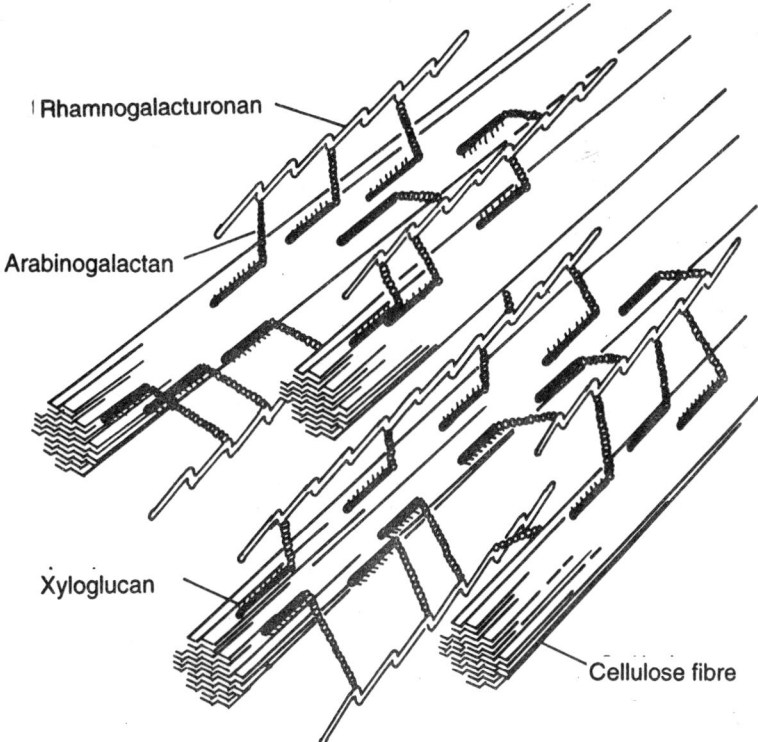

Fig. 3.1(e) Three dimensional model of the priimary cell wall - showing relationship between the aggregated cellullodr microfibrils and the hemicelluloces of the matrix

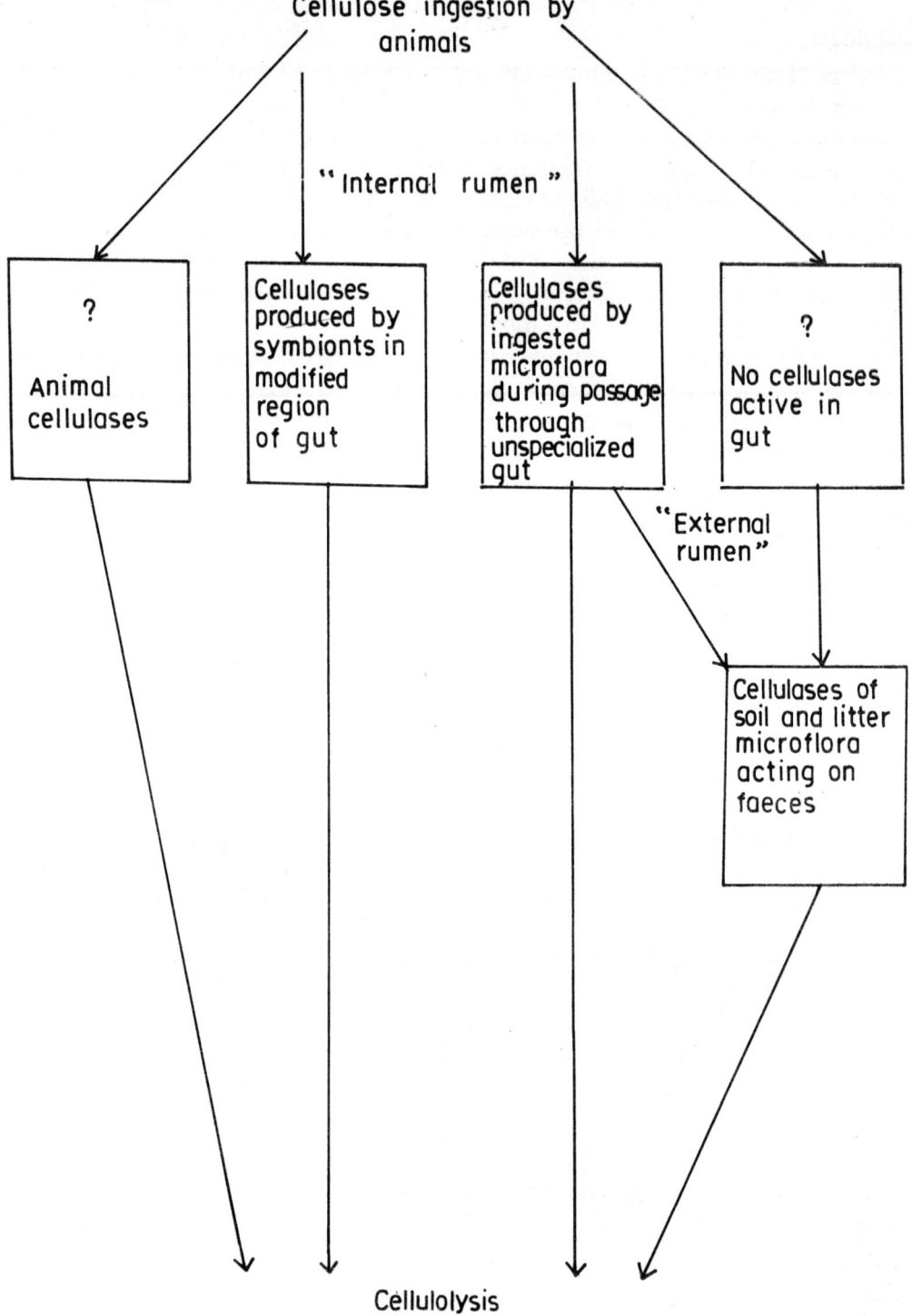

Fig. 3.1(f) The action and interaction of saprotrophic animals and microorganisms in cellulos decomposition

In higher plants the content of cellulose changes with the age and the type of plant. In the young stage of plant the cellulose content is usually low and it increases with the age of the plant. The amount is considerably high in woody structures. In grasses and legumes, generally the amount of cellulose is low and it accounts for as low as 15 percent of the dry weight.

Breakdown of cellulose is accomplished by a variety of microorganisms. Under diverse environmental conditions the process of cellulose degradation has been observed. The rate of decomposition may however, vary under different conditions. Cellulolytic microorganisms responsible for the decomposition of cellulose rich substrate are abundantly present in field and forest soils. Alexander (1961) remarked, "The physiological heterogeneity of the responsible microflora permits transformation to take place in habitats with or without O₂, at acid or at alkaline pH, low or high moisture levels and from temperature just above freezing to the extremes of the thermophilic range. The cellulose-utilizing population includes aerobic and anaerobic mesophilic bacteria, filamentous fungi, basidiomycetes, thermophilic bacteria, actinomycetes, and certain protozoa"

It has been generally observed that fungi are the main agents of cellulose degradation in humid soil where - as bacteria play a major role in semi arid conditions. Many fungi are responsible for cellulose decomposition. *Aspergillus, Trichoderma, Chaetomium, Curvularia, Fusarium, Phoma, Thielavia* and *Memnoniella* are strong cellulolytic fungi. These fungi are normally associated with softer tissues rich in cellulose whereas in the breakdown of woody tissues the significant contribution is by cellulolytic basidiomycetes.

Fungi being aerobes, decompose the cellulolytic material in the presence of enough oxygen. In addition to fungi, certain aerobic, mesophilic bacteria are also associated in the process of cellulose degradation. *Cytophaga, Sporocytophaga, Angiococcus* and *Polyangium* have been identified as cellulose decomposers. In addition to the above bacteria, species of Achromobacter, Pseudomonas, Vibrio, *Bacillus, Cellubrio* and *Cellfalcicula* are also listed by some workers to be associated with cellulose decomposition.

Among actinomycetes, species of *Streptomyces, Micromonospora, Streptosporangium* and *Nocardia* are cellulolytic. However, it has been observed that, though many actinomycetes have the necessary complement of enzymes their cellulolytic ability is low and they are much slower in attacking the polysaccharide than most fungi and true bacteria.

Certain protozoa are also capable of cellulose degradation. Tracey (1955) identified two genera - *Hartmanella* and *Schizopyrenus* for the process.

The decomposition of cellulose in the total absence of molecular oxygen is a common feature. During anaerobic conditions large quantities of ethanol and organic acids like acetic, lactic and butyric are produced due to anaerobic cleavage of the cellulose molecule. A number of bacterial species particularly spore-forming mesophiles, spore forming thermophiles, non-spore forming rod, cocci, several actinomycetes and also certain fungi may grow in the absence of

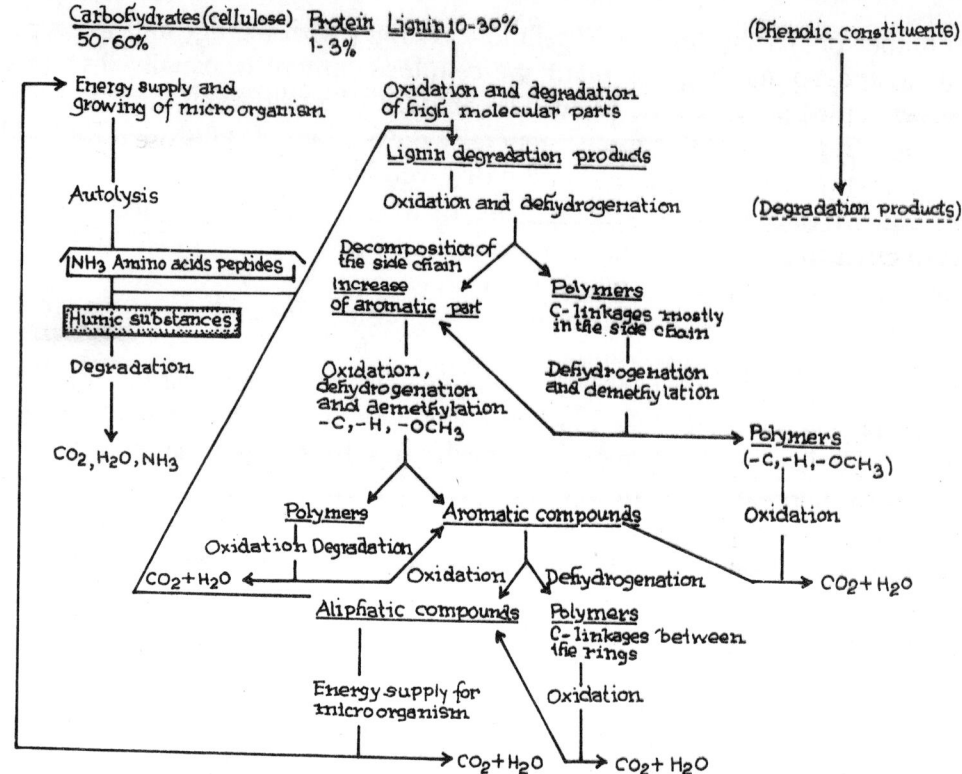

Fig. 3.1(g) Pathways of humus biosynthesis

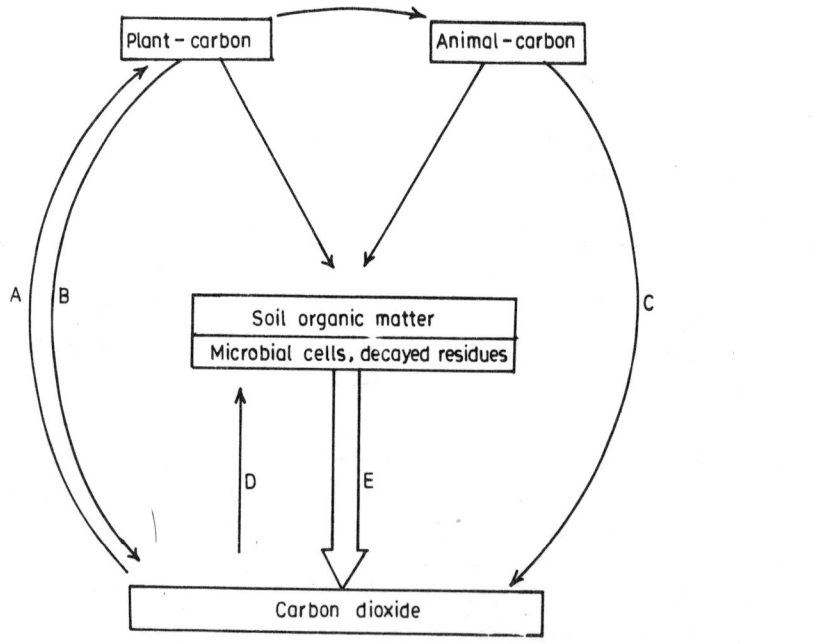

A. Photosynthesis. B. Respiration, plant. C. Respiration, animal. D. Autotrophic microorganisms.
Fig. 3.1 (h) The carbon cycle

molecular O_2 and they are associated with cellulose decomposition. *Clostridium*, a common soil dweller and also abundant in compost, manure, river mud and sewage is a good anaerobic cellulose fermenter. Fungi like Merulius and Fomes have the ability to survive in the absence of O_2 and they are reasonably good cellulolytic species. *Micromonospora* may be mentioned amongst the actinomycetes.

Degradation of cellulosic materials also occurs at relatively higher temperature. Both aerobic and anaerobic microorganisms are associated with the thermophilic transformation. The process appears to be in operation at as high as 61°C and degradation is quite fast at higher temperatures. *Clostridium thermocellum* and *C. thermocellulaseum* are the most active thermophilic bacteria. Both are obligate anaerobes and require a low oxidation - reduction potential for proliferation. Certain thermophilic fungi and actinomycetes also participate in the breakdown of cellulose at high temperatures.

Biochemistry of cellulose decomposition is different in case of aerobic and anaerobic conditions. In aerobic condition, CO_2 and cell-carbon are major products when the decomposition is accomplished by fungi and actinomycetes. In certain cases, small amount of organic acids is also released. Aerobic bacteria, on the other hand, produce only CO_2 and cell substances during the process. During anaerobic degradation however, the situation is a little different as far as the by-products are concerned. Both mesophilic and thermophilic anaerobic bacteria are incapable of metabolizing even simple substrates to completion. In such a situation, a large number of organic compounds are obtained as end products. CO_2, H_2, ethanol, acetic, formic, succinic, and lactic acids are generally the intermediate products.

Cellulose destruction is achieved by cellulase enzyme which catalyses the conversion of insoluble cellulose into simple, water soluble products. Mono-or disaccharides are liberated through cellulase catalysis. Subsequently the further steps involved in cellulose hydrolysis vary with respect to the individuals involved in the process. The aerobes convert simple sugars to CO_2 where as the anaerobes are responsible for organic acid and alcohol production.

Products of anaerobic decomposition of cellulose

Bacterium	*Products*
Mesophiles	
Clostridium cellobiopaiusm	CO_2, H_2, ethonol, acetic, lactic and formic acid
Clostridium dissolvens	CO_2, H_2, ethanol, acetic, latic and butyric acids
Bacteroides succinogenes	CO_2, acetic and succinic acids
Ruminococcus flavefaciens	Acetic, formic and succinic acids
Thermophiles	
Clostridum theromcellum	CO_2, H_2, ethonol, acetic, lactic, formic and succinic acids

Hemicellulose

Next to cellulose, hemicellulose is the major constituent of a plant. The hemicelluloses are water-insoluble polysaccarides and on hydrolysis they yield hexoses, pentoses and uronic acids. The latter, however, is not extracted from all the hemicelluloses. Based on the presence or absence of uronic acids hemicelluloses have been classified into two groups - Polyuronide hemicelluloses, those which have uronic acids and cellulases, without uronic acid.

Polyuronide hemicelluloses are intimately linked with other components of the cell wall and never occur in a free state. They are generally present in the form of lignin polysaccharide complexes. Chemical hydrolysis of polyuronide hemicelluloses yield two major polysaccharides both containing a pentose sugar and a uronic acid moiety. In one type, xylose and glucouronic acid are dominant and infact this type is most common in plants, where as in other arabinose and galacturonic acid are the major constituents.

On account of chemical heterogeneity, the decomposition of hemicellulose residues is of varied nature. Once the polysaccharide rich plant materials are added to the soil the rate of decomposition is very fast, which slows down later on. It thus appears that certain constituents decay fast whereas others take reasonably good time for their breakdown. By and large, however, the complete breakdown of the hemicelluloses in soil takes relatively longer time. Infact, it has been observed that at the initial stage the destruction of hemicelluloses is even faster than celluloses. As far back as in 1935, Acharya observed that in case of rice straw, hemicelluloses disappeared more rapidly under anaerobic conditions than cellulose. He noted that only 9.8 percent of the cellulose was metabolized after one month where as the percentage of hemicelluloses lost during the same period was 26.8. In aerobic conditions, he further observed that in the same period 62.4 percent of the hemicelluloses and 56.2 percent of the cellulose were mineralized.

As the chemical nature of hemicelluloses is very complex and of a varied type, the decomposition of this polysaccharide is also very interesting and complex. Many fungi, bacteria and actinomycetes are associated with the decomposition of hemicelluloses. A large number of microorganisms of the above category use the polysaccharide as a sole source of carbon and energy. Some common soil fungi like *Aspergillus, Penicillium, Fusarium, Trichoderma, Trichothecium, Rhizopus, Mucor, Humicola Cunninghamella, Zygorrhynchus, Chaetomium, Helminthosporium, Fomes* and *Polyporus* are capable of degrading one or the other type of hemicelluloses. Among bacteria, *Bacillus, Achromobacter, Pseudomonas, Cytophaga, Sporocytophaga, Lactobacillus* and *Vibrio* are important hemicellulose decomposers. *Streptomyces* has been observed to be associated with the process among *actinomycetes*. The microorganisms listed above form the major component of the soil microflora and the hemicellulolytic microflora is not restricted to a few genera or species, rather it includes a variety of fungi, bacteria and actinomycetes. At any given time in soil, one or the other form responsible for degradation of hemicelluloses are abundantly present thus resulting in the destruction of the different constituents of the polysaccharide. Also the destruction is accomplished under various soil conditions like aerobiosis, anaerobiosis, mesophilic, thermophilic etc. This gives an added advantage to hemicellulose decomposers.

Hemicellulases are the major enzymes for breaking down the various types of hemicellulose. The polysaccharides being relatively complex with high molecular weights need to be converted into simpler compounds before they can be utilized as a carbon source. With the help of extra-cellular enzymes the polysaccharide is hydrolysed to shorter and simple carbohydrate fragments. Hemicellulolytic microorganisms are responsible for initial hydrolysis. Once the complex hemicelluloses are hydrolysed, further breakdown of the resultant products is accomplished by another set of secondary population. Utimately simple sugar units are produced.

Lignin

After cellulose and hemicellulose, lignin forms the third major constituent of the plant tissue. Infact the amount of lignin in woody material is much higher. Lignin is present in the secondary layers of the cell wall and in the middle lamella. In young plants where the primary structures are more dominant, the lignin content is very low but it increases tremendously when the plant matures. Lignin is always associated with polysaccharides and never occurs in a free state. In woody tissues of plants, lignin content varies from 15 to 35 per cent where as in young tissues and grasses the amount is as low as 3 to 6 percent.

The lignin molecules are composed mainly of three elements carbon, hydrogen and oxygen. In contrast to cellulose and hemicelluloses the structure of lignin is aromatic. Roughly the percentage of carbon and hydrogen is 64 and 6. Methoxyl groups constitute 14 percent. These figures, however, are not fixed and the composition varies with the source and also sometimes the method of isolation. The basic unit in lignin probably is phenyl-propane (C_6-C_3) containing one methoxyl-carbon, C_6 representing the benzene ring linked to a C_3 propyl-type side chain. It is seen that one aromatic ring is found for each ten carbons. The benzene nucleus contains a hydroxyl group in addition to the methoxyl.

Like cellulose and hemicelluloses, the decomposition of lignin too proceeds either in the presence or in the absence of oxygen. The peculiar feature of lignin is its resistance to enzymatic degradation. Lignin being most complex and resistant to enzymatic activities, the pace of decomposition is very slow. In any given tissue, it is the lignin which is decomposed in the end. Once the cellulose and hemicelluloses are acted upon by the microorganisms and broken down, the plant material is usually left with lignin rich woody parts. It takes a pretty long time for the soil microorganisms to degrade the lignified materials. However, despite its resistance, the material is subjected to microbial attack and is broken down. The time span involved in the complete destruction of lignified tissues is much longer. It is a common sight to see the plant material lying in the soil for many many years, particularly those reasonably rich in lignin before they disappear due to microbial decomposition. In the forest, particularly, one will come across undecomposed lignin rich woody material for a number of years before they are completely degraded. During decomposition, it is thus clear, that the microflora consumes the individual organic components at different rates and in this process lignin is last to show appreciable oxidation.

The biochemistry of lignin degradation is little known. The Precise mechanism of enzymatic cleavage and the intermediate compounds formed in the process are not properly investigated. Generally the enzyme lignase is supposed to be involved in the break down process. The end-products formed are rapidly oxidized. It is believed that the intermediates formed are low molecular weight aromatic substances. The decomposition seems to proceed via the formation of vanillic acid which is ultimately converted to proto catechuic acid. It thus appears that lignin is probably depolymerized to simple aromatic substances such as vanillin and vanillic acid or possibly other methoxylated aromatic structures. One thing, is however, certain that enzyme system is extracellular. Ultimately the remaining methoxyls are removed with the formation of hydroxy-benzene derivaties which finally give low molecular weight organic acids.

Mainly the members of higher fungi i.e. of the basidiomycetes and a few ascomycetes are associated with lignin break down. Genera like *Agaricus, Coprinus, Armillaria, Fomes, Ganoderma, Lenzites, Marasminus, Pleurotus, Polyporus, Polystictus, Poria, Stereum, Ustulina, Clavaria, Clitocybe, Collybia, Lepiota, Mycena* and *Pholiota* are lignin decomposers. It is commonly observed that one or more of the above fungi are abundant on the decaying wood in the forests. Some of the above fungi are also cellulose decomposers. In one study there was observed that out of 46 soil inhabiting basidiomycetes 44 could degrade both lignin and cellulose. Species of *Clitocybe, Collybia, Mycena* and *Marasmium* may be included in this category. It has also been noticed that certain species of basidiomycetes have a special liking for the lignin and they are relatively poor in cellulose degradation. Surprisingly a few basidiomycetes are only lignin decomposers.

Role of bacteria in lignin decomposition is not very clear. However, it is assumed that certain bacteria must be capable of breaking down lignin, but their participation is certainly secondary to the fungi. It is suspected that in anaerobic destruction of lignified tissues, probably anaerobic bacteria play a greater role. Anaerobic decomposition of the lignins, however, is much slower than for cellulose and hemicelluloses. The biochemical pattern operating for the decomposition of the lignified tissue in vascular strands of leaves and decomposition of wood is probably also not the same.

At the stage where lignin begins to decompose, there is usually a marked increase in the amount of humic acids. It is also observed that during the microbial decomposition of litter, many of the fungal species which are present in greater abundance have dark pigments either in their hyphal walls or in their spores. Burges, Hurst and Walkden (1964) suggested that there is good reason to believe that humic acids are essentially complex polymers of phenolic materials and that the phenols concerned may have come from flavonoids present in the plant debris, from the decomposing lignin and perhaps also from microbial syntheses. Henderson (1947) opined that once lignin or humic acid has been degraded to simpler aromatic substances, these would form suitable substrates for a wide range of soil organisms. Degradation of a complex substrate like lignin would, however, involve a succession of microorganisms, each of which might carry out a relatively small number of biochemical steps.

Starch

Biodegradation of cellulose, hemicelluloses and lignins, the major constituents of plant tissues, has been discussed. Among the other constituents, starch occupies a prominent place in the plant materials. Infact, starch is second only to cellulose in plants. In photosynthetic plants, the material is accumulated in leaves in large amounts. It is the major reserve carbohydrate and in many species is stored in tuber, bulbs, corms, underground rhizomes, fruits and seeds. It also occurs in reasonably good quantity in xylem, phloem, cortex and pith regions. In addition to higher plants, microorganisms may also accumulate the polysaccharide.

Starch is mainly composed of two components, amylose and amylopectin. Amylose is characterised by having a linear structure built up of 200 to 500 or more glucose units which are linked together by an $\propto$-1, 4, glucosidic bonding. Amylopectin is also made up of glucose units which are linked together by $\propto$-1, 4 linkages but the molecule is branched and possesses side chains which are attached through $\propto$-1,6 glucosidic linkage. The 2 components of starch differ from each other both physically and chemically. The relative proportion of the two components varies in starch. Amylopectin normally occurs between 70 to 80 per cent, where as amylose is lesser and the percentage varies between 20 to 30.

The biodegradation of starch occurs both in the presence or absence of oxygen. Starch molecules being much more simpler in chemical nature to other components discussed earlier, degrades at a rapid pace. In plant tissues, infact it is the starch which disappears first during microbial destruction. The end-product of soil-enzyme hydrolysis is glucose. Maltose which is formed as an intermediate compound is rapidly hydrolysed yielding a simple sugar.

Breakdown of the starch molecules is accomplished by all the major soil microorganisms. Fungi, bacteria and actinomycetes actively participate in starch degradation. On account of diverse microbial population being associated with the process, the breakdown of starch proceeds in all diverse conditions. A large population of starch hydrolysers may be seen both in the rhizosphere region and also in the soil away from the root. Thus the starch decomposers are widely encountered in the soil.

A wide variety of bacterial species like gram positive and Gram negative Genera, spore formers, non-spore formers, aerobes and anaerobes are responsible for starch hydrolysis. Bacterial species of genera *Achromobacter, Bacillus, Chromobacterium, Clostridium, Cytophaga, Flavobacterium, Micrococcus, Pseudomonas, Sarcina* and *Seratia* have been reported to degrade starch. *Micromonospora, Nocardia* and *Streptomyces* among actinomycetes are good starch decomposers. In fact for most of the actinomycetes starch is a good source of carbon. With the exception of yeasts which are very poor starch decomposers, many filamentous fungi are endowed with suitable hydrolytic enzymes which catalyse the process of starch break-down. *Aspergillus, Fomes, Fusarium, Polyporus* and *Rhizopus* are a few filamentous fungi which need special mention for starch destruction.

Amylases are the major enzymes responsible for starch breakdown. Amylases are characteristically extra-cellular enzymes. ∝- and ß amylases are primarily concerned in the microbial breakdown of starch. The mode of action of the two enzymes varies to some extent. Both the enzymes though act upon the 1,4 linkage, the hydrolysis is retarded once the amylopectin branch point is encountered. It has been observed that ß-amylase is capable of acting upon both amylose and amylopectin. During the process, the enzyme cleaves every second glucose-glucose bond from the terminal end of the molecule. The enzyme ß-amylase is not capable of catalysing the hydrolysis of branch points of amytopectin. Through the enzymatic activity maltose fragments are liberated. ∝-amylase when acts upon the amylopectin, the end products formed have greater molecular weight than maltose. Finally however, the maltose is converted to glucose by mediation of the enzyme ∝-glucosidase. The process proceeds as given below:

$$\text{Starch} \xrightarrow{\text{amylase}} \text{Maltose} \xrightarrow{\text{ò-glucosidase}} \text{Glucose}$$
$$(\text{Glucose})^n \qquad\qquad (\text{Glucose})^2$$

Pectic

Pectic substances are present as calcium and magnesium pectates in the middle lamella of the plants. This tissue is located between the individual cells. Amount wise the pectic substances hardly constitute about one percent of the dry matter of plants. The pectic substances are complex polysaccharides. Galacturonic acid units linked to one another in a long chain compose the pectic substances. The carboxyl group of the galacturonic acid may be partially or completely esterified with methyl groups.

According to Alexander (1961) the followng three types of pectic substances are found in plants :

(a) Protopectin — A water insolube constituent
(b) Pectin — A water-soluble polymer of galacturonic acid with methyl ester linkages.
(c) Pectic acids — Water soluble galacturonic acid polymers which are devoid of methyl ester linkages.

The above three substances are closely related to each other.

Pectic substances are readily decomposed by soil microorganisms and all the three major constituents of soil living population i.e. fungi, bacteria and actinomycetes participate in the breakdown of pectic substances. During the process of decomposition the microorganisms derive carbon and energy sources for their survival and proliferation. A large number of pectolytic microorganisms are abundantly present in soil. Wieringa (1947) estimated the population of pectolytic organisms in soil and the estimate suggests that 10^5 to 10^6 pectolytic microorganisms are usually present in one gram of soil. The population of pectolytic microorganisms is considerably high near the root region and in many cases actinomycetes population is relatively higher than the other two groups i.e. bacteria and fungi. At higher pH regimes, where the pectolytic microorganisms are more, bacteria and actinomycetes are dominant. However, at lower pH level, fungi form the bulk of pectolytic microorganisms.

Three types of enzymes i.e. protopectinase, pectin methyl esterase (PME) and polygalacturonase (PG) are associated with the degradation of pectic substances. The activity of the three enzymes are different and they act upon different constituents of the pectic substances. Protopectinase is responsible for the decomposition of protopectin and soluble pectin is formed subsequently. PME hydrolyses the methyl ester linkage of pectin resulting in the production of pectic acid and methanol.

$$\text{Pectin} \xrightarrow{\text{PME}} \text{pectic acid + methanol.}$$

PG is associated with the breakdown of both pectin and pectic acid and destroys the linkage between galacturonic acid units. The end product of the break down is galacturonic acid.

Bacillus, Clostridium and *Pseudomonas* are the major pectin hydrolysers. *Bacillus* and *Clostridium* are responsible for retting of fibres in flax, hemp and jute. They digest the pectic substances of the middle lamella which results into loosening of the cellulose fibres from the residual tissues. *Fusarium* and *Verticillium* among fungi are the pioneer pectin decomposers. It is interesting to note that both the fungal species are root pathogens and cause wilts in plants. The enzymes released by the fungi catalyse the decomposition of pectic substances in the xylem cell walls. Pectin enzymes are responsible for solubilizing the middle lamella when they are released by pathogenic fungi and in the process the individual cells get separated from each other.

Inulin

Inulin is also a polysaccharide and is made-up of fructose units which are known as fructosan. 25 to 28 fructose residues in the carbohydrate chain bound in 1,2 linkage form the inulin molecule. It is a storage carbohydrate and is found in tubers, roots, stems and leaves.

A number of microorganisms are reported to hydrolyse inulin. Two enzymes fructonases and inulinase are reported to activate the degradation of the polysaccharide. Inulinase, an extracellular enzyme has been reported to be highly active and this is found in many fungi. *Pseudomonas, Flavobacterium, Beneckea, Micrococcus, Cytophaga* and *Clostridium* are the bacteria reported to be inulin decomposers. Certain actinomycetes too are associated with the breakdown of inulin. *Streptomyces* is one of the most active actinomycetes. Among fungi *Penicillium funiculosum*, *Aspergillus fumigatus* and *Fusarium moniliforme* are the important genera to decompose inulin. Fructose polymers, levanbiose or free fructose are generally the end product of the degradation of inulin.

Gums

Normally two types of gums are found – gum arabic and mesquite gum. Arabinose, rhamnose, galactose and glucouronic acid are present in gum arabic whereas arabinose, galactose and methyluronic acid are the main constituents of mesquite gum. Gums are exuded usually from the different parts of the plant particularly from bark, leaves and roots. Mostly the higher plants are responsible for the formation of gums. *Acacia spp.* yield gum arabic and gum mesquite is obtained from *Prosopis juliflora*.

Besides higher plants, a number of microorganisms are also associated with the formation of gums. Gums and related substances are synthesized by certain bacterial species. *Rhizobium spp.* are mostly responsible for the extracellular gum synthesis. Similarly Bacillus spp. also produce gum. The chemical nature of the gums produced by the two bacteria differs. Glucose, pentose and glucouronic acids are present in the gum synthesized by *Rhizobium*, whereas *Bacillus* gum contains fructose or glucose. A few other bacteria which synthesize gum yield arabinose and galactose upon acid hydrolysis.

The decomposition of gums is brought about by a variety of microorganisms. Aerobic and anaerobic mesophiles and thermophiles are usually responsible for the breakdown of gums. Normally bacterial species like *Pseudomonas*, *Cytophaga*, *Achromobacter* and spore forming bacteria are involved in the destruction of gums. Basidiomycetes among fungi are major gum decomposers. Certain bacterial species are associated with the digestion of bacterial gums. Gram negative spore forming *Bacillus spp.* are of special significance in this connection.

Chitin

The sources of origin of chitin in soil are mostly the insects and fungi that live in the soil. Cell walls of basidiomycetes and yeasts are usually rich in chitin. Chitin is also found in the skeletons of many invertebrate animals. In filamentous fungi the amount of chitin varies between 2.6 to 26.2 percent on a dry weight basis.

Structurally chitin is a polysaccharide and the basic unit is an amino sugar. A long chain of N-acetylglucosamine units arranged linearly constitute chitin.

The chitinoclastic population in soil is largely made-up of actinomycetes. It has been observed that approximately 90-99 percent of the chitinoclastic population in soil is composed of actinomycetes. Bacteria and fungi usually play little role in the breakdown of chitin rich materials. The digestion of chitin takes place normally in aerobic condition. In well aerated soil the population of chitin digesters may be as high as 700 million per gram. *Streptomyces* and *Nocardia* have been reported to be the main bacterial species for the digestion of chitin. Vedkamp (1955) observed that in less aerated, poorly drained habitats, bacteria are the main chitin decomposers. Bacterial species like *Achromobacter, Bacillus,* *Beneckea, Cytophaga, Chromobacterium, Flavobacterium, Micrococcus* and *Pseudomonas* are associated with the transformation of the polysaccharide. Vedkamp (1955) further observed that certain non-spore forming bacteria possibly related to coryneform metabolize chitin. Besides, several species of basidiomycetes, *Fusarium, Mucor, Trichoderma, Aspergillus, Mortierella, Gliocladium, Penicillium,* *Thamnidium* and *Absidia* are the main fungi involved in chitin hydrolysis.

A number of substances have been reported to be liberated during the break-down of chitin. Presence of N-acetylglucosamine, glucosamine, acetic acid and ammonia are usually the end products (Vedkamp, 1955). On further metaboliztion the end products yield CO_2 and cell protoplasm. In most of the chitin decomposers - extracellular enzyme chitinase is present and this enzyme is supposed to be responsible for catalysis in the process of chitin digestion. Chitinase is supposed to degrade chitin to the disaccharide stage and another enzyme, a chitobiase further acts upon the disaccharide to convert it into two acetylglucosamine fragments.

From the above discussion it is clear that in the soil a large variety of different substrates are available in varying amounts for microbial decay. The nature and amount of the materials available for a given soil at a time differs depending upon the availability of the litter. There also exists a wide heterogenity in the microbial populations which are involved in the breakdown of the available material for decay. The biochemical processes involved in the decomposition of the various substrates, too, differ considerably. Readily soluble substances such as simple sugars, and amino acids appear to be absorbed directly and then metabolized within the microbial tissue. Other relatively complex substances, such as starch, pectin etc. are broken down by extra-cellular enzymes which can at times diffuse appreciable distances from the organisms which produce them and hydrolyse their substrate. The products of hydrolyses are then available for absorption not only by the organisms which cause the initial hydrolysis but by any other microorganism which may be in the near vicinity.

It is also evident from the above discussion that microbial utilization of the sugars, starches and pectins seems to be very rapid and it is difficult to detect any appreciable quantities of these in the litter. The condition is also different when the decomposing ability of the microorganisms is tested in laboratory and field conditions. Many microorganisms which prove to be efficient decomposers in the laboratory, fail to do the job in natural environments. It is interesting to note that mycorrhizal fungi which are so conspicuously associated with roots are devoid of the ability to decompose organic substrate in the root region.

The process of decomposition is a very complex one. Complexity exists both in the nature of chemical substances present in any substrate and also the microorganisms involved during the decomposition. A simple leaf may take reasonably good time for breakdown and a number of microorganisms may be associated in the process. The whole process of microbial decay of the substrates is very systematic and well regulated. Simpler chemical moities are first to decompose and the process continues to more and more complex constituents. In the process, a succession of microorganisms appear and disappear. This continues till the last traces of the litter is lost in the soil and the chemicals thus released are incorporated in the body of the microorganisms. The microorganisms have enough patience to wait for their turn to come during the process.

Besides microorganisms, certain soil animals like earthworms etc. also help in the breakdown of plant litter. Infact the incorporation of fall materials into the soil, which is normally a rapid process is brought about by earthworms and other animals. The fallen plant materials may be cut up and removed into the burrows or smaller leaves may be dragged down whole. Such smaller fragments of the substrate are then exposed to microbial decay.

Factors affecting microbial decay in soil

The rate at which the different substances are metabolized in soil is governed by a number of environmental factors. Soils as such varying in their physical and chemical characteristics exert considerable influence upon the process. In addition to the soil the major environmental factors which influence the transformation are the available nitrogen level, temperature, aeration, moisture, pH, the relative presence of components in the substrate. The biochemical heterogeneity of the substrate and the nature and density of the microorganisms present in the vicinity of the decaying substrates are the other factors affecting the decomposition. Placement of the litter in the soil horizon is also to be taken into consideration during the biodegradation process.

As has been discussed in chapter 1, the well aerated poros clayey soil is usually rich in microflora. The litter added to such soil are subjected to rapid decomposition on account of rich microbial population. More compact soil or poor in gaseous exchange is not very conducive for litter decay. More the microbial population, faster is the rate of decomposition. In many cases the application of inorganic nitrogen to the soil enhances the break down of litter in soil. Ammonium or nitrate salts usually are suitable sources. Within a reasonable limit, the rate of decomposition is proportional to the concentration of nitrogen added. Manure and organic nitrogen compounds such as urea, amino acid, peptone and casein also enhance the conversion rate. Alexander (1961) observed that approximately 1 unit of nitrogen is required for every 35 units of cellulose oxidised. Further 3 parts of nitrogen are incorporated into microbial protoplasm for 100 parts of cellulose decomposed.

Bio-conversion of the litter proceeds at a wide range of temperature. As has been discussed in the preceeding pages, the breakdown of litter is accomplished from freezing temperature to 65-70°C. Different types of microorganisms i.e. psychrophilic, mesophilic and thermophilic are equally responsible for microbial decay of litter. This is one of the reason why at any place on the earth, the decay and transformation of the litter continues. Depending upon the location, different types of microflora are usually available for the breakdown. The rate of decomposition, however, varies under different temperature regimes. In most cases the rate of turnover of the litter is maximum between 20- 30°C and in that case is accomplished by mesophiles. But a number of thermophilic fungi, bacteria and actinomycetes are in no case less efficient than their counter part mesophiles. Rather sometimes the thermophilic species are more efficient. This is the reason, for the rapid decomposition of compost where the temperature inside the compost heap sometimes rises as high as 55-60°C. It is not uncommon to see the rapid degradation of litter in hot springs where also the temperature is fairly high.

Like temperature, aeration too is important for microbial activity, though here too extreme situations are noticed. As discussed, almost in every case, the decay proceeds both in presence and absence of oxygen. Aerobic and anaerobic decay of litter is a common happening. I have enumerated earlier many micro- organisms which are active in both the conditions for microbial degradation. Normally since the aerobic species of the microbes are more in number, it is obviously safer to conclude that the rate of decay is more and faster in aerobic environment. This is the reason for the quick decomposition of litter at the upper few inches of soil. However, the role of anaerobic bacteria, actinomycetes and a few fungi has not to be underestimated. Along with aerobic forms, the anaerobes, however small their population may be, have their own role to play. At extremely low temperatures too certain bacteria are equally active.

Moisture is one of the most important attributes for microbial activity. At moderate moisture levels, the conditions are conducive for the growth of most of the soil microorganisms which at one stage or other are responsible for the breakdown of litter. Extreme moisture regime, low or high are not suitable for microbial decay of litter. At low moisture levels, the normal activity and growth of the soil microbes are hampered. The microbes, then simply survive by producing different resting bodies and under such conditions the whole process of decomposition slows down. In tropical countries, during summer months when the soil moisture drops down to reasonably low levels, particularly at the upper horizon of soil where litter deposition is maximum, litter breakdown is almost negligible. However, as soon as the rains start, there is a rapid influx in the microbial activity. Irrespective of high temperature, in the beginning of the rainy season, availability of moisture stimulates microbial decomposition. In very much wet conditions, when the soil becomes flooded and microbial growth slows down due to poor aeration, the rate of decay is again affected adversely.

Like other soil parameters, pH of the soil also has wide range in relation to microbial growth and proliferation and consequently litter breakdown. In neutral to alkaline environments, many microorganisms grow and liberate appropriate enzymes for the hydrolysis of various substances. Actinomycetes and bacteria as discussed earlier are more active in this regime and we now know quite clearly the significance of these two groups of microflora in litter breakdown. By and large, however, fungi prefer acidic side of the reaction. It thus appears that due to the active involvement of different soil microorganisms over a wide range of pH, the process of decomposition continues at a fairly large pH range. At the acidic side, fungi are more active, whereas towards neutrality and alkaline side bacteria and actinomycetes play the major role. The different chemical constituents of the litter are of significance in microbial breakdown. Simpler substances like sugars, starch, amino acids etc. are hydrolysed quickly. Complex constituents starting from cellulose to lignin and chitin, are broken-down later on. On analysis of litter on the soil, it is seen that leaves are therefore first to decay and woody material the last. In any mixed substrate, it is easy to observe the leaf to disappear from the scene, followed by simple primary twigs etc. and the final that is stock which is left rich in lignified tissues takes years for final disposition. However, the story is a never ending one because the fall of litter is a continuous process. In nature, if the soil profile is examined regularly, it shows that the top of the soil always receives new litter in varying amounts. As one proceeds deeper and deeper, litter in different stages of decomposition is encountered.

Finally the type of the microorganisms and their population at the site of decomposition are the most important factors to reckon with. It is very important that the site in question where the litter breakdown has to be accomplished must be equipped with suitable microflora. It is not merely their presence which is important but the population has also to be considered. Unless suitable microflora with reasonably good density is available the whole process will have its own fate accomplii. The microorganisms must also be in an active stage of growth with efficient enzymatic backup to catalyse the process of microbial decomposition.

Looking at all these aspects, I stated earlier that the process of microbial decomposition is very complex in every respect but is an extremely important one. If I say, no decomposition, then no release of nutrient, no cycling of the minerals and finally no survival and growth of life on the earth - I will not be incorrect.

Chapter 4

Microbial decomposition of herbicides

The microbial breakdown of natural substances present in plant and animal tissues has been elaborated in chapter 3. Besides these natural substances, soil also receives the large number of synthetic materials which are in use now to control plant pathogens and weeds. In advanced countries the use of such chemicals is a common practice and a large amount of chemicals are dumped into the soil in the course of their application to plants or the soil. To control plants from plant and animal pathogens repeated application of fungicides, nematocides, insecticides is a common feature. In crop fields many undesirable weeds grow and they affect the growth and yield of crops. The amount of loss incurred due to the presence of weeds, varies depending upon the type and intensity of the weed. The emergence period of the weed is also responsible for the damage done to crops. To get rid of such unwanted plant species from the field, different types of chemicals known as herbicides are sprayed over crops. The substances thus applied irrespective of their mode of application, are ultimately added to the soil in varying amounts. The chemicals applied to plants are not retained on the plant surface in totality, a considerable amount of them is generally added to the soil. It is difficult to avoid this inspite of the best precautions taken. Besides this the plant remains of such chemical-treated plants are also incorporated into the soil as litter. Thus the soil, with repeated incorporation of chemicals, gets loaded with undesirable harmful chemicals. I say undesirable and harmful because such substances once accumulated and mixed with soil have a harmful effect on the soil. The normal structure of soil is changed as the chemicals are normally toxic once their concentration rises above certain limits.

Collectively such chemical substances are grouped as pesticides. Pesticides are substances designated to control or eradicate a specific pest of economic crops. Pesticides include four major groups of substances i.e. herbicides, insecticides, fungicides and nematocides. As pointed out earlier, the use of such chemicals is more frequent in advanced countries. In developing countries like ours the problem is not that serious because due to financial constraints such chemicals are not that frequently used in our fields. Farming in our country is largely the monopoly of uneducated people who are involved in the agriculture profession. Due to ignorance about the use of pesticides, coupled with the economic limitations, our agricultural fields are luckily still free from such hazardous menaces. However, this escape is not likely to last long because slowly farmers are getting involved in more modern and advanced types of farming so as to reap better yields. With this in mind the use of pesticides is now gaining momentum, and in the next few decades it is quite possible that our lands will also have the same problem which the fields of advanced countries are facing.

The wide spread use of synthetic organic substances for combating insect pests, plant pathogenic fungi or nematodes and thus improving crop yields, may be regarded as a recent phenomenon. The discovery of the insecticidal properties of dichloro-diphenyl trichlorethane (DDT) in 1939 by the Geigy chemical company together with the recognition of the value of 2-4- dichlorophenoxy-acetic acid (2,4-D) as a selective herbicide against dicotyledonous weeds during the course of world war II, by among others, Nutman, Thornton and Quastel (1945) marked the beginning of this modern era of plant protection. Some others also subsequently noticed the limited persistence of 2,4-D in unsterilized soil and also initiated studies on microbial degradation of herbicides and pesticides.

As discussed earlier, soils are a complex mixture of weathered minerals, plant and animal organic debris in different stages of decay and they contain a wide variety of plants, animals and microorganisms. Soils are exposed to many physical, chemical or climatic influences, including varying moisture content, temperature fluctuations, radiation, wind, erosion etc. Consequently, chemical substances added to soil are subjected similarly to a variety of chemical, physical and biological forces, and they may react or be degraded in many ways. In addition to biological forces, the substances added to soil are also subjected to other forces like leaching, volatiliztion, chemical oxidations, and photochemical reactions for their degradation. Plant and animal processes also operate in the soil ecosystem simultaneously. The main concern here, however, is the microbial degradation of such added substances.

A large number of substances have also been identified as potential herbicides. Phenoxyalkyl carbonic acids, substituted ureas, nitrophenol, chlorinated acetic and propionic acids, phenyl carbamates, thiolcarbamates and some others have been widely used to control specific weeds. It will be out of the scope of this book to discuss in detail the biodegradation of all the chemicals. The discussion in this chapter will centre around a few selected ones.

2,4-D

Destruction of 2,4-D takes about 2-8 weeks or more depending upon the dose used for application. Also the physical and chemical environments of the de-toxifying microbial population, which is rarely the same for different soils, plays an important role. Many bacterial species are responsible for metabolizing 2,4-d. Species of *Achromobacter*, *Arthrobacter*, *Corynebacterium*, *Flavobacterium* and *Mycoplasma* are associated with the break-down of dichlrophenoxyacetic acid. It has been observed that for complete mineralization, removal of the side chain, dehalogenation and cleavage of the aromatic ring are necessary steps. When a dilute solution of 2,4-D percolates through soil, three phases are noticed. Initially, a slight fall in 2,4-D concentration is observed which is due to adsorption of a small amount on soil. This phase is followed by a long period with scarcely any change in concentration and this is referred to as the lag period during which either as a result of mutation or by adaptation, a population of microorganisms able to metabolize the 2,4-D begins to develop. The third phase involves a rapid logarithmic break-down of 2,4-D, coincident with the growth of 2,4-D adapted microbial population, which ceases only when the 2,4-D supply is exhausted. During the subsequent addition of 2,4-D, however, the first and second phases are normally curtailed and with the development of suitable microorganisms in the environment, 2,4-D is readily degraded. Certain factors, however, like temperature, pH, moisture condition, aeration and substrate or other nutrient supplies which favour aerobic bacterial growth, influence the degradation process. As a rule, any management practice which favours the microbial proliferation in soil leads to a more rapid metabolism of 2,4-D. Thus it is obvious that operations like improving the moisture status, raising the temperature and liming shorten the period of effectiveness of 2,4-D and many other herbicides (Hernadez and Warren, 1950; Newman and Norman, 1947).

2,4,5 trichlorophenoxyacetic acid, the other important herbicide, differs structurally from 2,4-D in possessing a third chlorine atom on the benzene ring. Structurally thus there is no remarkable difference between 2,4-D and 2,4,5-T but the latter is more persistent in the soil and may be detected in non-sterile soil for a period of six months to a year after its application (De-Rose and Newman, 1947). Not much is known about the microbial degradation of this compound.

Normally the rate of decomposition is faster at the upper horizon of the soil and neutral soils are more conducive for the break-down of herbicides. During the process of microbial decay of herbicides, it has been observed that certain growth stimulating substances are formed as intermediate compounds which in turn influence the growth of plants as well as underground microorganisms. On the contrary, phytotoxic intermediates may also be formed during the decomposition of herbicides. Thus, the process of herbicidal break-down in the soil can swing the balance in either direction and may provide a fascinating field of investigation.

Phenylcarbamate, phenylurea and acylanilide are used for plant protection purposes. Most of such compounds are derived from different substituted anilines. It has been observed that most of these compounds are degradable in soil by microorganisms. The degradative pathway may vary; parent anilines are liberated during the break-down. Compounds like monuron (3-(p-calorophenyl)-1,1-dimethylurea), linuron, a chloro derivative of monuron have been seen to be degraded in soil. Certain bacteria have been reported to degrade aniline and its chloro-derivatives. Biodegradation of propham, propanial solan and Swep has been reported by Kaufman and Blake (1973). *Pseudomonas, Achromobacter* among bacteria and fungi like *Aspergillus, Penicillium* and *Fusarium* have been accounted for the break-down. *Lipomyces starkeyi*, a yeast species degrades parquet (1,1-dimethyl-4,4-bi- pyridylium-2A). *Pseudomonas* spp. are also reported to metabolize Parquat. Another related compound Diquat is also easily degraded by soil bacteria. Often a particular organism may only be able to perform one step in the breakdown process and complete removal from soil depends on a series of organisms. A persistent chemical may be altered but not degraded and the product may be just as difficult to degrade and sometimes as toxic as the original compound. Though a compound may not be degraded when presented as a sole carbon source, it is often broken down (C0-metabolized) when in the presence of utilizable carbon, which is the situation in most environments.

It has been observed that fungicides O-tolua-nilide, and 2,5- dimethyl-furan-3 carboxyanilide and herbicides isoprophyl-n- phenyl-carbamate are all meta-bolized. Decomposition of several phenylurea herbicides by *Bacillus sphaericus* - a bacterium present in soil, has been reported by Wallnofer and other (1971). They noted that *B. sphaericus* hydrolyses have acylamide group, affording the corresponding anilines and acid; 2-methyl and 2-chloro-benzonic acid anilines and the various N'- methoxyphenylureas. Monolinuron, linuron, metabromuron and melaran are all degraded. The pathway followed for break-down is

$$R_1 - \langle \underline{\quad} \rangle - NHCOR_3 + H_2 \longrightarrow R_1 - \langle \underline{\quad} \rangle - NH_2 + R_3COOH$$
$$\quad R_2 \qquad\qquad\qquad\qquad\qquad R_2$$

Where R_1, R_2 = Cl : R_3 = NH.Alk,O. Alk, Alkayl group.

Other organophosphorus insecticides are subjected to microbial decomposi-tion. Malathion may be degraded to various intermediates by *Rhizobium spp.* or by *Trichoderma viride* (Mostafa *et. al.*, 1972). These insecticides are much less persistent than the very stable organochlorine compounds.

Most of the compounds discussed in this chapter do not occur naturally; some are toxic to microorganisms, and are resistant to decomposition. Persistent chemicals present the problem that they accumulate in organisms near the top of food- chains to a much higher level than in the environment in general. Most of the biocides have, however, been noticed to be biodegradable and most of them do not last more than a year in the soil system. Bacteria and fungi are usually responsible for their break-down. Chemicals which are usually resistant include those with substituted amino, methoxy, sulphonate and nitro- groups, chlorine substitution, particularly in the metaposition, metasubstituted beneze rings in general, ether linkages and branched carbon chains. It has also been noted that in such cases, small chemical changes can have a pronounced effect on degradability. Generally, the decomposition or immobilization of a compound in the environment is also affected by the rates of application and the proprietary formulation.

The adsorption of many pesticides to soil particles and the organisms which degrade them poses another problem. In such a situation, the rates of break-down depend on whether the active sites of the enzymes and/or the appropriate groupings on the molecule are exposed after adsorption. The degradation of foreign synthetic compounds that are added to soil, depends on the induction of suitable enzyme systems in the potential microorganisms which are expected to degrade the compound. The microorganisms have to adapt themselves to decompose synthetic materials a task for which they are ill equipped in normal circumstances. Once adapted, such microbes become more efficient and have an ecological advantage over the other microorganisms in the system, using the foreign molecules as a carbon and energy source. The adapted microbial populations then grow, multiply and decompose foreign materials in stages, releasing CO_2 and water as end products. Spokes and Walker (1974) observed that different genera of soil bacteria when adapted to grow on phenol or benzoate, can oxidise various chlorophenols or chlorobenzoates to a chlorocatechol. *Bacillus* was noticed to oxidise 3-chlorobenzoate to 3-chloro-2, 3-dihydroxybenzoate. Once such reactions take place the product becomes prone to futher oxidation. In a nut shell, it may be concluded that even partial oxidation is a step in the direction of the eventual detoxification of persistent chlorinated aromatic substances.

It is commonly observed that once foreign synthetic materials have been added to the soil system, initially the biodegradation of such compounds is a problem. This is because in the soil, there are no suitable microorganisms with efficient enzyme systems to act upon such materials. Slowly, however, microbes adapt and the enzymes, through various mechanisms as stated earlier, develop to catalyse the compounds. Once such a new set up develops in the microbial population, further addition of synthetic compounds does not pose much of a problem. They are degraded at much faster rates. Nevertheless, caution has to be taken while using such compounds because ultimately there is a limit to which the microbes can help in the process of degradation. If the concentration of such compounds crosses the threshold level, soil becomes toxic and in such a case it affects the microbial composition of the soil which is by and large stable for any system. Once this happens, the natural soil gets modified and the growth of plants is affected. Inspite of the versatile bio-chemical activities of the microbial world, it has been noticed that many organic molecules are completely resistant to their attack. This fact has to always be taken into consideration when such synthetic substances are used.

Chapter 5

Transformation of minerals

In chapters 3 and 4 the biodegradation of natural and synthetic substances, which are from time to time added to the soils, has been discussed. In addition to such substance there are other mineral elements like phosphorus, sulphur, calcium, potassium and iron etc. which play an important role in the growth and metabolism of all organisms. Microorganisms also require these minerals in varying amounts for their survival. Microorganisms contain them and therefore immobilise them and these minerals are lateron released as a consequence of the decay of organic matter. The products of microbial metabolism by changing pH, for example, may affect the availability of an ion by altering its solubility or its oxidation state, though that ionic species is not itself metabolized. Sulphur and iron, on the other hand, are used as substrates by certain bacterial species. In this chapter the microbial conversion of phosphorus, sulphur, calcium, potassium and iron will be considered.

Transformation of phosphorus

Phosphorus is essential for all life and is present in phospholipids, nucleic acids and ATP. The amount of phosphorus needed by living organisms is much smaller in comparison to carbon and nitrogen. However, many a time, it may be an important limiting nutrient.

Phosphorus exists in the environment in the form of phosphate, usually orthophosphate, on a very small scale, the presence of phosphine (PH_3), hydrogen phosphide (P_4H_2), phosphite ($PO_3"$) etc. has also been reported. This is usually a result of microbial oxidation and reduction of phosphorus. The amount of phosphorus in soil is reasonably high and it varies between 400 to 1200 mg kg^{-1}. Usually it occurs as insoluble inorganic phosphates, particularly of calcium and iron and in organic complexes- specially inositol phosphate, but in such a state it is hardly ever available. Plants and microorganisms too enrich the level of phosphorus in soils after their death. Another source of phosphorus input is from fertilizers and a slow addition from rocks. The key processes in the phosphorus cycle are thus the decay of organic matter and the dissolving of inorganic phosphorus. The decay of organic matter is accomplished by a wide variety of saprophytes. As phosphorus is usually present as phosphate, it is merely released by the breakdown of the carbon skeleton.

Both plants and microorganisms need phosphorus in reasonably good quantity, its requirement being second only to nitrogen. During cellular metabolism phosphorus is required in certain essential steps of the accumulation and release of energy.

According to Alexander (1961), microorganisms are responsible for a number of transformation of the element. These include :

(a) altering the solubility of inorganic compounds of phosphorus;

(b) mineralizing organic compounds with the release of orthophosphate;

(c) converting the inorganic, available anion into cell protoplasm, an immobilization process analogous to that occurring with nitrogen; and

(d) bringing about an oxidation or reduction of inorganic phosphorus compounds.

Microbial mineralization and immobilization reactions are particularly important to the phosphorus cycle in nature. The two processes are to a great extent responsible for the availability of phosphorus.

The vast quantity of vegetation that is continuously incorporated in soil and undergoes decay forms the chief source of organic phosphorus compounds for soil. Normally the amount of phosphorus in the tissues of agricultural crops varies between 0.05 to 0.50 percent. Phytin, phospholipids, nucleic acids and nucleoproteins, phosphorylated sugars, co-enzymes and related compounds are the major phosphorus rich constituents in plant tissues. Sometimes, vacuoles and internal buffers may also contain inorganic orthophosphate. In the plant tissues, phosphate is not reduced, rather it enters organic combinations largely in an unaltered state. Phosphorus consequently is present in plants and microorganisms in different combinations. In phytin, phospholipids and nucleic acids it is present in the form of phosphate. A calcium-magnesium salt of phytic acid constitutes phytin. A combination of phosphate with lipids yields phospholipids. Lecithin is made of glycerol fatty acid, phosphate and choline. Proteins and nucleic acids form nucleoproteins.

In soil, phosphorus is largely (25% to 85%) present in an organic form. Surface vegetation, microbial population or the metabolic products of microorganisms form the bulk of organic phosphorus. The upper layer of soil is much richer in organic phosphorus content than the sub-surface horizons. As insoluble calcium, iron or aluminium phosphates it occurs in inorganic forms. In the form of calcium salts the concentration of phosphorus is more in neutral or alkaline soils, whereas in acidic soils, the iron and aluminium salts predominate. In upper horizon of soil, where humus formation is more, phosphorus is present in the form of inositol phosphates which originate from phytin and nucleic acids or in their degradation products. Though present in reasonably high quantity in the form of phytin or related substances in soil, phytin is mineralized very slowly and consequently is not beneficial to plants. In acid soils particularly, the rate of phytin mineralization is very slow. In alkaline soils, however, the reaction is fast and phytin phosphorus is then made available to plants. It is also interesting to note that inositol phosphates react readily with iron, aluminium and magnesium salts and in such a situation the solubility of the salts is very poor.

The presence of phosphorus in soil in the form of nucleic acids or nucleotide derivatives is not of much significance. In most soils, nucleic acid type compounds probably contribute less than one to a maximum of 10 percent of total organic phosphorus. Nucleic acids, primarily RNA, are bound with in the living cells of microorganisms and they are mineralized only after the death of the microbial cells.

It has been observed that soil rich in organic matter are correspondingly rich in organic phosphorus. Though organic phosphorus is plentifully available inorganic matter rich soils, it is unexpectedly not readily mineralized, though microbial enzymes should degrade them at a rapid rate. The cause for slow decomposition is the adsorption of the phosphate containing substrates. Certain clay minerals are active in the adsorbent process. Also, acidic soils between the pH range of 3.5 to 4.5 are more conducive for the adsorption of phytin. Neutral and alkaline soil, on the other hand, are far better where little adsorption takes place. Some phosphorus is immobilized by the microorganisms causing decay : fungi contain 0.5 to 1.9 percent (by weight) and bacteria 1.5 to 2.5 percent. The amount in higher plants is in the range of 0.5 percent to 5.0 percent. There is, therefore plenty of phosphorus in most decaying substrates and the concentration needs to be greater than about 0.2 percent before some release occurs.

A large amount of phosphorus, generally present in form of inorganic compounds is not available to plants because it is in an immobile state. A large number of microorganisms present in soil have the unique quality of bringing the immobile phosphorus into soluble form. This feature is a common phenomenon in a large population of soil microflora. It has been observed that about one tenth to one half of the bacterial species present in soil have the quality to solubilize calcium phosphate. Species of *Pseudomonas*, *Mycobacterium*, *Micrococcus* and *Flavobacterium* among bacteria and *Penicillium*, *Sclerotium*, *Aspergillus* and many other fungi are active in the conversion.

During the conversion process, a part of the phosphorus is assimilated by microorganisms but infact the amount made soluble and released is in excess to the requirement of the microorganisms. The excess amount thus released is made available to plants. In this conversion process, organic acids play an important role. Nitric and sulphuric acids are also of significance. These organic and inorganic acids convert $Ca_3(PO_4)_2$ to the di or monobasic phosphates and the latter are then readily made available to the plants. The amount of carbohydrates being oxidized by heterotrophs has a great impact upon the solubilization process. Transformation proceeds rapidly if enough of carbonaceous substrate is converted to organic acids. Nitric and sulphuric acid which are produced by the oxidation of nitrogenous materials or inorganic compounds of sulphur, act upon rock phosphate. During this process phosphate solubilization is affected. Liberation of soluble phosphorus from rock phosphate composts is facilitated by nitrification of ammonium salts. Generally the solubilization of phosphate is achieved by acid production. Mobilization of ferric phosphate is on the other hand accomplished by certain bacteria which liberate hydrogen sulfide. Sperber (1957) observed that hydrogen sulphide reacts with ferric phosphate to yield ferrous sulphide and phosphate is liberated. Phosphate is also released from iron and aluminium phosphate in oxygen deficient conditions.

The presence of a large number of phosphate dissolving microorganisms in soil is well known. It is also an established fact now, that the root region is abundantly rich in phosphate dissolving microorganisms. On account of this, the phosphate assimilation by higher plants is enhanced.

The role of mycorrhizal fungi in uptake and translocation is of considerable importance to plants and this aspect has already been elaborated in chapter 2 of the book. It is a well established fact that uptake of phosphorus is increased in soil of moderate fertility. The beneficial effect of suitable mycorrhizal fungi is profitable to young trees or their seed beds with suitable fungus/fungi, before planting new plants (Jha *et. al.,* 1991).

It has been noticed that many bacteria, fungi and actinomycetes are responsible for the release of bound phosphorus in crop residues and soil organic matter which is ultimately available to plants. Both mesophilic and thermophilic microorganisms actively participate in the mineralization of phosphorus. Warm temperatures usually favour decomposition and due to this thermophilic species have a dominant role to play. Like wise pH of soil is also important in the process. Nucleic acids are dephosphorylated most readily, followed by lecithin and phytin is the slowest to be degraded. Normally, the release of phosphate is relatively rapid under conditions favouring ammonification. Phosphorus, present in bacterial cells is mineralized at a faster rate. Acid soluble organic phosphorus, phospholipid and DNA are dephosphorylated in a short time. On the other hand phosphorus of microbial RNA is released more slowly.

The enzyme phytase is responsible for the liberation of phosphate from lytic acid or its calcium magnesium salts, phytin and subsequently for the accumulation of inositol. Species of *Aspergillus, Penicillium, Rhizophus, Cunnighamella, Arthrobacter* and *Bacillus*, which are abundant in soil, synthesize the enzyme. The population of phytase rich microorganisms is generally enhanced with the addition of carbonaceous material to the soil.

Like nitrogen, phosphorus is both mineralized and immobilized in soil. Both the process operate in soil and are governed by the amount of phosphorus in the plant residues undergoing decomposition and the nutrients required for the associated microbial population. In case the concentration is higher than that needed for microbial nutrition, the extra amount accumulates as inorganic phosphate. If the amount on the other hand is less for the microorganisms, it gets immobilized. This fact has a wide repercussion in soil, because a part of the available nutrient supply is immobilized if the substances undergoing decomposition are poor in phosphorus or have a wide C:P ratio. Phosphate is formed when the ratio narrows with time due to CO_2 volatization.

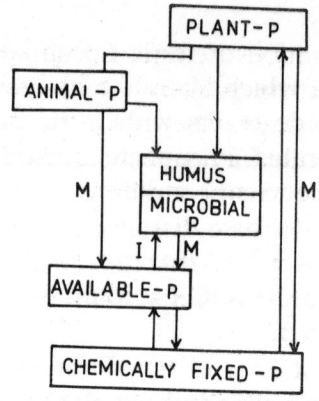

Fig. 5.1(a) The Phosphorus Cycle
M. Mineralization
I. Immobilization

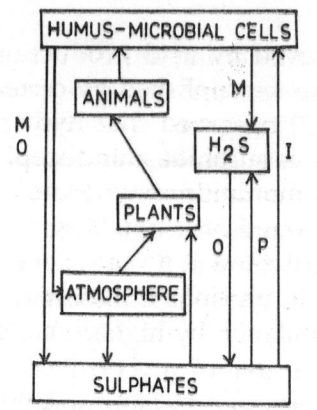

(b) The Sulphur Cycle
M-Mineralization, I-Immobilization,
O-Oxidation, R-Reduction

Transformation of sulphur

Soil normally contains 10 to 15 percent sulphur in the form of sulphate. Sulphur is also released from organic matter during its decomposition. To a great extent higher plants are dependent on sulphur which is released during microbial mineralization. Sulphur is also added to soil in the form of that released to the atmosphere during the burning of fossil fuels, which is then washed down by rain to finally reach soil and water. Through the above sources the total amount of sulphur input to the soil is estimated to be usually 100-500 mg kg^{-1}, of which half may be in an insoluble form. The presence of sulphides and polysulphides has frequently been noticed in anaerobic soils. In unpollulated air the amount of sulphur, mostly in the form of sulphate, sulphur dioxide and hydrogen sulphide, is much less approximately 1 to 5 ng l^{-1}. In polluted conditions values upto 200 ng l^{-1} have been reported.

In natural conditions, sulphur is usually not a limiting factor. However, in agricultural fields where intensive cropping is done and high doses of nitrogen fertilizers are applied, it may be necessary at times to add additional sulphur to the soil. Sulphur both in organic and inorganic forms, is readily metabolized in soil. The microorganisms responsible for the conversion of sulphur are affected by environmental factors and as a consequance of this the rate of mineralization of the element is also affected. Alexander (1961) suggested the following four different ways by which the process of sulphur mineralization is accomplished :

(a) decomposition of organic sulphur compounds, a process in which large molecules are cleaved to smaller units and the latter in turn converted to inorganic compounds;

(b) microbial assimilation or immobilization of simple compounds of sulphur and their incorporation into bacterial, fungal or actinomycetes cells;

(c) oxidation of inorganic compounds such as sulphides, thiosulphate, polythionates, and elemental sulphur; and

(d) reduction of sulphate and other anions to sulphide.

Plants and animal tissues are reasonably rich in sulphur and once they die and the remains incorporated into soil, the proteins of tissues are hydrolysed by microorganisms to the amino acid stage. The latter, through microbial attack, release sulphate and sulphide which gets accumulated in soil. The fate of sulphur compounds is governed by the presence or absence of O_2 in environment. In aerated soils, sulphur is metabolized to sulphate, in anaerobic or water-logged situations it accumulates in the form of hydrogen sulphide (H_2S). A number of intermediate compounds are formed before the final storage of sulphide and sulphate, however, the concentration of such intermediate compounds in nature is usually very low.

In soils, sulphur is largely (10-15%) present in the form of sulphate and unless it is adsorbed to soil colloids it is readily leached. The remaining 85-90% of sulphur is locked up in organic matter which is released slowly during the decomposition of litter. In addition to organic matter, sulphur also enters soil in the form of animal wastes, chemical fertilizers and rain water. Most of the sulphur in soil is in organic combination, and the inorganic sulphate concentration is very low. In organic fraction, sulphur is associated with amino acids like cystine, cysteine and methionine. Sulphur is also found in the form of ethereal sulphates, thiourea, glucosides and alkaloids in organic matter. Sulphur may be found in many forms like elemental sulphur, sulphide, sulphur in amino acids and sulphate. Coversion of such sulphur materials is carried out by microorganisms. Most sulphur conversions seem to involve adenose-S-phosphosulphate (APS) and parts of the cytochrome system (cytochrome and cytochrome C_3) with NAD, NADP or Ferredoxin. Reduction also involves 3 phospho adenosine-S-phosphosulphate (PAPS). The major steps are as follows :

$$ATP + SO_4'' \rightleftharpoons APS + PO_4''$$
$$ATP + APS \rightleftharpoons PAPS + ADP$$
$$PAPS + NAD\,(P)\,H \rightleftharpoons PAP + SO_3'' + NADP\,(P)^+$$
$$NADH + SO_3 \rightleftharpoons SH' + NAD^+$$

Dissimilatory reduction (bacteria) :

$$ATP + SO_4'' \rightleftharpoons APS + PO_4''$$
$$APS + 2e' \rightleftharpoons SO_3 + AMP$$

Oxidation (bacteria)

$$2AMP + 2SO_3'' \rightleftharpoons 2APS + 4e'$$
$$2APS + PO_4'' \rightleftharpoons 2ADP + 2SO_4''$$

Organic sulphur is metabolized during decomposition of protein by microbes and depending upon aeration conditions, is released either as sulphate or hydrogen sulphide. During anaerobic conditions volatile organic sulphur compounds are produced. A part of the sulphur available is used by microorganisms and is thus immobilized. Under continued anaerobiosis sulphides may be oxidized by photosynthetic bacteria, whereas in aerobic conditions chemosynthetic bacteria are responsible for oxidation. Some hetero-trophic bacteria, actinomycetes and fungi also have the capability to oxidize hydrogen sulphide. The major role in the oxidation of elemental sulphur is played by the chemosynthetic Thiobacillus spp. Besides elemental sulphur, sulphides, thiosulphate and tetrathionate are also oxidized to sulphate.

The activities of *Thiobacillus* have great impact on pH of the environment. Due to the production of acid solubility is increased, which facilitates the availability of other elements. In soils rich in organic matter, photosynthetic bacteria like *Chlorobium* and *Chromantium* are of more importance.

Mineralization of sulphur is a relatively complex process because it is present in various forms. Besides plant and animal tissues where it is abundantly present, it is also associated with microbial proteins, amino acids cystine and methionine, B vitamins, thiamine, biotin and thioctic acid. It also occurs in the tissues and excretory products of animals as free sulphate, taurine and to some extent as thiosulphate and thiocyanate. It has been observed that sulphides are among the major inorganic substances released during the decomposition of proteinaceous substrates. As stated above mineralization of sulphur is accomplished both in the presence and absence of atmospheric oxygen. Many bacteria have the ability to form H_2S from partially degraded proteins. As such, during decomposition proteinaceous substrates sulphides are the major inorganic substances released. Mineralization of many other sulphur rich compounds like cystine, cysteine and methionine proceeds in an entirely different manner. The microorganisms involved in the processes are also of diverse types. In well aerated soils, cystine and cysteine are converted to sulphate. In addition to bacteria, fungi like *Microsporeum gypseum* also convert cysteine sulphur to sulphate. This process proceeds through the formation of cystine, sulphinic acids, sulphite and finally to sulphate (Stahl *et. al.*, 1949).

In soil, methionine is usually more resistant to microbial attack. Volatile compounds are produced during the transformation of methionine in soil. Many a time it is difficult to detect such volatile compounds. Mineralization of thiamine and taurocholate is quite rapid. On the contrary, substances like thiourea, mercaptoethanol etc. are slowly oxidized.

Normally it has been noted that C:S ratio of the decomposing substrate and sulphur content govern the extent of mineral sulphur formation. Microbial need is another important parameter which regulates the accumulation of sulphate. If the amount of sulphate is more than that of the microbial requirement, it gets accumulated in the soil. The rate of sulphur mineralization is also affected by environmental factors which in turn are responsible for microbial growth. The sulphur requirement of microorganisms is of a varied nature. Sulphate, hyposulphite, sulphoxylate, thiosulphate, persulphate, sulphide, elemental sulphur, sulphite, tetrathionate and thiocynate of inorganic substances and cysteine, cystine, methionine, taurine and undecomposed proteins of the organic group are certain important sources of sulphur for microorganisms.

Transformation of Iron

Iron is needed by microorganisms in a very small quantity. In nature, particularly in terrestrial systems, iron is present in reasonably good quantities and it is a major constituent of the earths crust. Though present in good quantities the element is not in an available form for plant utilization. It is not uncommon to note that many plants suffer due to iron deficiency. Iron is thus needed for the proper growth of plants, including microorganisms.

Transformation of iron, like other elements, is accomplished by various microorganisms. Depending upon the type of microorganisms involved in the process, the mechanisms are also different.

Alexander (1961) suggested the following biological means for the transformation of iron.

(a) A group of bacteria known as iron bacteria oxidise ferrous iron to the ferric state which precipitates around cells as ferric hydroxide.

(b) Soluble organic iron salts are acted upon by many heterotrophic microorganisms and the iron is converted into a slightly soluble inorganic form which is subsequently precipitated.

(c) Insoluble ferric ions are converted to soluble ferrous form by microoranisms through a change in the oxidation - reduction potential. With the growth of microorganisms a decrease in oxidation - reduction potential is noted in the surrounding, which facilitates the conversion.

(d) Iron is also brought into solution with the help of many acids such as carbonic, nitric, sulphuric and organic acids. A number of bacteria and fungi produce such acids as end products responsible for the process.

(e) In the absence of oxygen, sulphide is formed from sulphate and organic sulphur compounds, which remove iron from solution as ferrous sulphide.

(f) Soluble organic complexes are also formed by microorganisms through the liberation of certain organic acids.

Mainly two groups of microorganisms are associated with the conversion of iron. Release of iron during the decay of organic matter is mediated by heterotrophs. The second group is autotrophic bacteria which derive energy from the oxidation of ferrous iron to the ferric state. In organic matter iron is present in varying amounts in cytochromes. Depending upon the pH of the environment and microorganisms involved, iron is released during the decay of organic matter. It may be either in soluble form, immobilized or may be precipitated. Certain chelating agents are produced by some organisms which convert iron to the soluble form and later make it available to the needy organisms. During microbial metabolism certain acids are released, making the environment acidic which helps in the release of iron from the insoluble inorganic state. Solubilization of iron may also occur under reduced conditions. Ferrous iron normally is more abundant in solution form when the soil condition is below pH 5, where as ferric iron is favoured above pH 6.

A variety of microorganisms are responsible for the transformation of iron in nature. Such microorganisms are commonly found in soil, water and in iron pipes. Bacteria associated with iron transformations are known collectively as iron bacteria. Such bacteria belong to various taxonomic groups. The genera of iron bacteria are distributed amongst four families namely Caulobacteriaceae, Siderocapsaceae, Chlamydobacteriaceae and Crenothrichaceae. Alexander (1961) placed the iron depositing bacteria in three divisions :

(a)   obligate chemoautotrophs - bacteria which utilize ferrous oxidation as the sole energy - yielding reaction for growth;

(b)   facultative chemoautotrophs - such bacteria normally use organic matter and oxidise it, they are also capable of utilizing ferrous salts as an energy source; and

(c)   heterotrophic organisms - bacteria which accumulate ferric hydroxide. They, however, do not obtain energy from iron oxidation.

Sartory and Meyers (1948) suggested three bacterial genera namely *Gallionella ferruginea*, *Thiobacillus ferrooxidans* and *Ferrobacillus ferrooxidans* to be obligate iron autotrophs. *Thiobacillus ferrooxidans* is interesting in that it has a very low pH optimum i.e. 1.7 to 3.5. *Thiobacillus*, *Pedomicrobium* and *Sphaerotilus* deposit iron in their cell walls or extra cellular polysaccharide and they are supposed to be important in the geochemistry of iron. Besides the above, *Pseudomonas* and *Aspergillus* too have been noted to deposit iron extra cellularly.

In soil, the population of typical iron bacteria is not of significance because they are found in very low quantity. More significant in soil are the heterotrophic species. In soil, iron associated with certain water soluble organic compounds is of more importance and is infact responsible for altering the availability of the element. Iron present in organic materials provides energy for the growth of microorganisms and is released and precipitated as insoluble ferric salts. Strains of diverse genera and families are responsible for the removal of iron from solution by attacking the organic portion of the salts. Crawford (1956) and Lewis (1928) observed that among bacteria *Aerobacter*, *Pseudomonas*, *Serratia* and *Corynebacterium*, many filamentous fungi, *Nocardia* and *Streptomyces* are active in the conversion.

Micro-biological ferric reduction is achieved through several mechanisms. Fermentable substrates also play an equally important role in reduction. Mobilization of iron is favoured in a high acidic environment. During microbial metabolism when O_2 level is lowered, the condition is favourable towards the reduction of ferric salts. Direct reaction of fermentation products with ferric hydroxides and oxides also leads to reduction. *Bacillus polymyxa*, *Bacillus circulans*, *Escherichia freundii* and *Aerobacter aerogenes* are usually associated with the conversion of ferric to ferrous iron.

In anaerobic conditions, species like *Desulphovibrio* are more active in the transformation of iron. In this process, sulphide is produced either through organic sulphur mineralization or by the reduction of sulphate. Hydrogen sulphide formed by microorganisms, leads to the precipitation of iron as ferrous sulphide by a reaction of H_2S with iron salts.

The process of iron transformation has a special significance with respect to iron and steel material buried underground. In the deep layers of soils, where anaerobic conditions prevail, corrosion of metals is a common phenomenon. The anaerobic microorganisms act upon metals and make the iron pipes useless after a few years thus leading to great economic loss. Microorganisms involved in the destruction of iron pipes in soil are more active at moderate temperatures, above pH 5.5, less free oxygen and in the presence of sulphate.

Mineralization of Manganese

Besides the elements discussed in preceding pages, transformation of the manganese is also equally important. This is required by plants in very small quantity and is included as an essential micronutrient for overall growth and development. In soil, Mn, is present in tetravalent form as the exchangeable divalent manganous ion. Normally Mn is taken up by plants in the divalent manganous form, Mn^{++++} is insoluble where as Mn^{++} is water soluble. pH of the environment plays an important role in the ionic set up of the element. Above pH 5.5 Mn is present in the form of exchangeable Mn^{++}. Above pH 8.00 Mn^{++} is oxidised to manganic oxides. In such a situation plants are incapable of assimilating manganic oxides and they suffer from deficiencies of the element. Autooxidation of Mn^{++} is a low hydrogen ion concentration process.

$$MnO_2 + 4H^+ + 2e^{--} \xrightarrow[\text{alkali}]{\text{acid}} Mn^{++} + 2H_2O$$

Microorganisms usually play an important role between pH 5.5 and 8.00. A large number of microorganisms are involved in the oxidation of Mn. The most active microorganisms are *Aerobacter, Bacillus, Corynebacterium* and *Pseudomonas* amongst bacteria and *Cladosporium, Curvularia, Helminthosporium* and *Cephalosporium* from fungi. In addition to these, certain *Nocardia* and *Streptomyces* also are capable of oxidizing the element. Alexander (1961) suggested that oxidation of the element is favoured more in dual cultures of microorganisms than the pure one. He observed that pure cultures of *Corynebacterium sp.* and a *Chromobacterium* are incapable of manganous oxidation, whereas a mixture of the two is able to carry out the transformation. The exact reason for the phenomenon is difficult to explain.

There is a close similarity between the transformation of iron and manganese. Many organisms which precipitate ferric hydrate are also capable of accumulating MnO_2. The latter compound induces a dark zone around cells of microorganisms. Deposition of iron or manganic oxides is widely reported in sheaths of *Leptothrix spp., Crenothrix polyspora* and *Clonothrix putealis*.

The population of Mn-oxidizers varies for different soils. Usually the population of such microorganisms is high in the root region. In soil Mn is present mostly in the form of MnO_2 or similar compounds. However, it is not uncommon to observe the presence of Mn_2O_3 and Mn_3O_4 also in soils.

For the oxidation of Mn, different mechanisms have been proposed. Oxidation of Mn ions has been noticed in the presence of hydroxy acid. During decomposition of carbohydrate rich materials, citrate tartrate, lactate, malate or gluconate are produced. This is more common in alkaline conditions and in such a situation the oxidation of Mn ions is also noticed. Microorganisms thus help the oxidation process either by producing hydroxy acids or by decreasing the hydrogen ion concentrations or by both. Kenten and Mann (1952) suggested another mechanisms in which enzymes play an important role. Particularly those enzymes which produce H_2O_2 i.e. enzyme peroxidase, bring about an oxidation of Mn ions.

The level of exchangeable Mn is increased during microbial metabolism. It is a common phenomenon that during microbial metabolism pH decreases, oxidation - reduction potential is lowered and the O_2 amount is at a low level. These factors favour the level of exchangeable Mn. Flooding of soil, coupled with the available carbohydrates and/or plant tissues, stimulates the soluble divalent cation.

It may be concluded that Mn is usually present in divalent, tetravalent and other oxidation states of the element and microorganisms which synthesize acid during microbial metabolism are largely responsible for regulating Mn availability. Acidity, the population, the presence of O_2 and the type and quality of organic matter present determine the form in which Mn irons are available in soil.

Chapter 6

Nitrogen cycle

Next to carbon, nitrogen forms one of the important constituents of plant and animal requirement. It is an important and essential constituent of all cells. In plants the amount of nitrogen varies from 1 to 10% by weight, whereas in animals the amount is between 20-30%. It is normally found in cells in the form of organic molecules, specially proteins.

In nature, nitrogen in gaseous form is abundantly present in the atmosphere, about 80% of air by volume. It is also available in geochemical deposits. In the latter case nitrogen is found as ammonium or occasionally nitrate in various rocks. It is a pity that although in the gaseous form nitrogen is plentifully available in nature but it is not taken by the plants in gaseous form and in plant nutrition nitrogen is assimilated almost entirely in the inorganic state as nitrate or ammonium. In the organic state nitrogen is associated with organic matter and with many nitrogen rich plant residues but in such a condition it is not available to the plants. Mineralization of nitrogen i.e. conversion of organic nitrogen to the inorganic state is essential for plant nutrition. During the process of mineralization, ammonium and nitrate compounds are accumulated and organic nitrogen disappears. Microorganisms are responsible for the mineralization and the two processes, ammonification and nitrification are involved. During ammonification, ammonium is formed from organic compounds and in nitrification, the ammonium is oxidized to nitrate.

As stated nitrogen in soil is present in the form of various nitrogenous organic combinations like humus, proteins, nucleic acids and related compounds. Although a variety of heterogenous microbial population is responsible for the mineralization of organic nitrogen compounds the process in general is relatively slow. Nitrogen rich organic compounds are resistant to microbial attack and only a small portion of the nitrogen reservoir of soil is mineralized during a particular time. Various hypotheses have been proposed for the slow mineralization of organic nitrogen. Two most commonly accepted ones are described.

According to one hypothesis, complex substances like lignin, protein are formed by protein and non-nitrogenous components of humus and in such a condition proteins become relatively less susceptible to digestion. The second hypothesis states that proteins are entrapped within the lattice of clay crystals and get resistant to microbial attack. The proteolytic enzymes are also adsorbed by clay and become less active in their function.

Role of microorganisms in the nitrogen cycle :

Microorganisms play an important role in the nitrogen cycle. Numerous bacteria and fungi are responsible for the decomposition of proteins, chitin, urea and other nitrogenous complex materials. Bacteria and fungi are the major components for chemical conversions between the various oxidation states of nitrogen. The latter includes ammonium, nitrogen gas, nitrite and nitrate. Soluble amino acids, small peptides or proteins or insoluble proteins and chitin are generally the nitrogen compounds available for microbial decomposition. It has generally been noticed that during decomposition nitrogen is almost always in short supply. Plant residues have C/N ratio 30 : 1 or higher and protein is about 5 : 1. Under the circumstances organic nitrogen in the residue is repeatedly recycled by microbes. A part is hydrolysed to amino acids and is used by microorganisms and the rest is deaminated to yield ammonia and carbon is released through respiration. In natural environments, existing organic materials are the major source of nitrogen for plants by mineralization.

Various processes are involved in the oxidation of nitrogen. The main steps are:

1. Nitrification — Ammonium is oxidized to nitrite and then to nitrate.
2. Nitrate reduction — Nitrate is reduced to nitrite and then to ammonium.
3. Denitrification — It is a reduction process and usually nitrite is reduced to gaseous nitrogen.
4. Nitrogen fixation — Dinitrogen gas is reduced to ammonium.

The amount of available nitrogen in an ecosystem is regulated by the above processes. The balance between the above four processes is largely controlled by aeration.

Nitrification

This is a very important process in the nitrogen cycle. In this process the conversion of ammonium to nitrate takes place. By and large two important steps are involved, first the production of ammonium as a result of decay of nitrogenous organic matter. This step is accomplished mainly by *Nitrosomonas* which converts ammonium to nitrite. In the second step the nitrite is converted to nitrate by another bacterium - *Nitrobacter*. Some other bacteria i.e. *Nitrocystis* and *Nitrosococcus* are also responsible for the process in aquatic systems, particularly is seas. Ammonium produced is taken up directly by microorganisms and some higher plants. For higher plants, however, nitrate is often a more favoured nitrogen source.

The conversion of nitrite to nitrate is a very fast process and at any given time the level of nitrite in the system is very low. The organisms responsible for the process are obligate aerobes and chemoautotrophs. They derive energy from

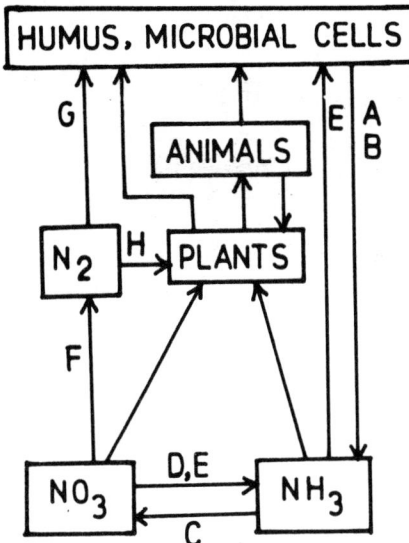

The nitrogen cycle
A. Ammonification
B. Mineralization
C. Nitrification
D. Nitrate reduction
E. Nitrogen immobilization
F. Denitrification
G. Non-symbiotic N_2 fixation
H. Symbiotic N_2 fixation

(a) The Nitrogen Cycle

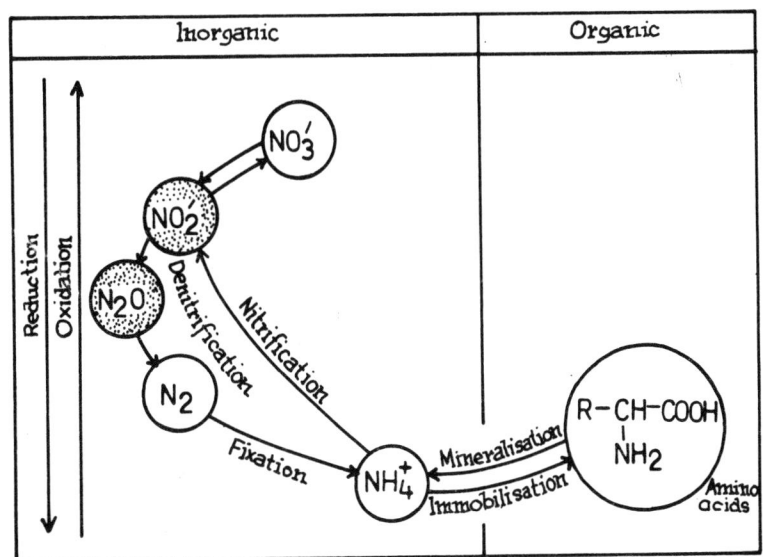

(b) Miniralization and Immobilization of nitrogen

Fig. 6.1

oxidation and can use this to fix carbon. During the conversion of ammonium to nitrite 65-66 Kcal Mol^{-1} energy is made available. Like wise in the conversion of nitrite to nitrate about 17 -18 Kcal mol^{-1} energy is obtained. In all aerobic environments nitrification is accomplished by chemoautotrophs. The complete process of nitrification has not been properly investigated from the chemical point of view. There are certain intermediate compounds which are suspected to be formed during the process. The first step is apparently the oxidation of ammonium to hydroxylamine using molecular oxygen. Hydroxylamine as such is toxic but its concentration is usually very low. There is always a shift towards ammonium and the reaction proceeds only as the hydroxylamine is oxidized to nitrate. Campbell (1983) suggested the existence of certain intermediates, possibly the nitroxyl radicle and nitrohydroxylamine. However, the free existence of such intermediates has been a question of uncertainty. Parts of flavoprotein and cytochrome system are needed for the oxidation of hydroxylamine. *Nitrobacter* ultimately oxidizes the nitrite to nitrate. This is achieved in a single step with molecular oxygen as the terminal acceptor, mediated by a cytochrome system, and ATP is generated.

The nitrifying autotrophs are all obligate and they depend upon inorganic materials for energy. The oxidative capacity of such autotrophs is limited solely to nitrogen compounds. For their cell synthesis carbon is derived from CO_2, carbonates or bicarbonates. The energy for the reduction of CO_2 is obtained by the oxidation of inorganic nitrogen compounds. All polysaccharides, structural constituent, amino acids and vitamins found in nitrifying autotrophs are synthesized from inorganic nutrients and CO_2. Not only this, such bacteria are also capable of synthesizing all enzymes and other factors needed for life from inorganic materials. Both in *Nitrosomonas* and *Nitrobacter* the heterotrophic stage in their life cycle has not been reported so far.

The process of nitrification is favoured near neutrality of the soil reaction. It has been observed that the strains of *Nitrosomonas* and *Nitrobacter* are very sensitive with respect to pH. The nitrifying isolates do not carry out nitrification much below pH 6.0. It is surprising to note that looking into the significance of nitrate production for plant growth, chemoautotrophic nitrifying genera are few in number. Though *Nitrosomonas* and *Nitrobacter* are relatively common in nature, other nitrifying bacteria like *Nitrosococcus* and *Nitrosospira* are less prevalent. The population of autotrophs is larger in soils of pH greater than about 6. It has also been observed that species of *Nitrosomonas* and *Nitrobacter* are normally closely associated in their occurrence. This is of great significance because the association helps in regulating the level of nitrite, which if accumulated higher than a certain level, is phytotoxic. Environmental conditions have a great impact on the occurrence and population of the nitrifiers. Warmer temperatures with enough humidity are more conducive for the growth of the autotrophs. On the contrary, drying or freezing decreases their abundance. These conditions, however, never y eliminate these bacteria.

mally, in most soils, nitrite does not accumulate. However, in alkaline soils, nitrate production from $(NH_4)_2SO_4$ is suppressed and this usually happens when the rate of ammonium application is increased. Alkalinity and high ammonium levels result in nitrite accumulation.

Nitrate reduction

Reduction of nitrate through nitrite to ammonia, is a process opposite to nitrification. Reduction is carried out intracellularly by a set of microorganisms. This is accomplished by certain bacteria, fungi and cyanobacteria which use nitrate and then reduce it to the amino group ($-NH_2$) in proteins. Usually nitrite and hydroxylamine are the intermediates. It proceeds under aerobic conditions and the enzyme is repressed by free ammonium.

In the absence of O_2 certain bacteria like *Thiobacillus denitrificans* and some species of *Pseudomonas, Bacillus, Micrococcus, Achromobacter* and a few fungi use nitrate as a terminal electron acceptor. In such a situation dissimilatory conversion of nitrate, particularly in soil or water is mostly accomplished by facultative anaerobes. In water-logged soils and deep waters anaerobic microhabitats develop due to rapid decomposition of organic matter and such situations are conducive for the above process to proceed. Nitrate reductase, an enzyme which contains molybdenum to acts as catalyst in the reaction.

Denitrification

The process involves the microbial reduction of nitrate and nitrite which ultimately results in the liberation of molecular nitrogen and in a few cases nitrous oxide. Nitrogen is thus lost to the atmosphere and transformed into unavailable forms.

It has been suggested that the volatization of nitrogen is achieved by following four reactions :
(a) Non-biological losses of ammonia;
(b) Chemical decomposition of nitrite under acid conditions to yield nitrogen oxide;
(c) Production of nitrogen by the non-enzymatic reaction of nitrous acid with ammonium or amino acids; and
(d) Microbial denitrification leading to the liberation of nitrogen and N_2O.

Sometimes the volatization of free ammonia is quite high particularly above pH 8.0. Like wise warmer conditions are also quite favourable for the process. In acidic conditions, nitrite decomposes rapidly to nitric oxide and NO. pH below 5.5 is generally favourable for the reaction. Though nitrite is produced by biological means either through nitrification of ammonium or by reduction of nitrate, its decomposition is purely a chemical phenomenon. pH of the environment has special significance in the volatization of nitrogen. Below pH 5.5, when oxidative or reductive processes are in operation and nitrite is unstable, the volatization of nitrogen takes place and NO is produced. Considerable loss has also been noted in acid soils where the pH is 5.5 or below. In such a situation it is mostly on account of nitrification.

Molecular nitrogen is also released through the chemical reaction of nitrous acid with amino acids or ammonium salts.

$$RNH_2 + HNO_2 \longrightarrow N_2 + ROH + H_2O$$

Microbial denitrification, is however, a major means of nitrogen volatization and during the process N_2 and N_2O are evolved. Normally, however, the amount of N_2 is much higher than N_2O. In certain acid surroundings, NO is also evolved during denitrification, however, the amount of this gas is too low under normal circumstances. The oxide is formed by the acid-dependent decomposition of nitrite produced by the reduction of nitrate and it has a close relationship with pH.

$$3HNO_2 \longrightarrow 2NO + HNO_3 + H_2O$$

The NO thus formed is reduced to N_2 by microorganisms. It may also get oxidized in air to nitrogen dioxide, NO_2.

Role of microorganisms

A large number of microorganisms, mainly bacteria, are involved in denitrification. In aerable soils, the population of denitrifying microorganisms is usually very high. The population may be in excess of a million per gram of soil. It has also been observed that the population of denitrifiers is much higher in the root region of plants. In this region, therefore, if conditions are conducive for organisms to change from aerobic respiration to a denitrifying type of metabolism, the rate of volatization is supposed to be very high. The active denitrifiers are species of *Pseudomonas, Achromobacter, Bacillus,* and *Micrococcus.* Species of *Pseudomonas* and *Achromobacter* present quite frequently are dominant in various soils. Besides the above bacteria, *Thiobacillus denitrificans, Chromobacterium, Mycoplana* serratia and *Vibrio* species also catalyse the reduction.

By and large denitrifying bacteria are aerobic. However, they are capable of using nitrate as the electron acceptor for their growth and survival when the O_2 supply is poor or absent. Presence of oxygen or nitrogen therefore determines the nature and activity of denitrifiers as being aerobic or anaerobic respectively.

Reduction of nitrate is not the monopoly of denitrifiers only. Many other microorganisms are also capable of reducing nitrate. Such conditions are observed during the process of protein synthesis or as a substitute for the reduction of O_2 in conventional aerobic metabolism. In protein synthesis, the microorganisms capable of using nitrate as a nitrogenous nutrient, reduce it to ammonium which helps in the process. Under anaerobic conditions, in presence of nitrate, the typically aerobic bacteria, stimulate the formation of nitrite or ammonium. Normally nitrite thus formed disappears rapidly but certain strains accumulate nitrite.

Alexander (1961) suggested three microbiological reactions of nitrate:

(a)   a complete reduction to ammonium, frequently with the transitory appearance of nitrite;

(b)   an incomplete reduction and accumulation of nitrite;

(c)   a reduction to nitrite followed by the evolution of gaseous compound i.e. denitrification.

Microorganisms that use nitrate as a nitrogen source carry out reaction "a" others which are incapable of complete reduction are to be supplied with ammonium or other reduced nitrogen compounds for their growth.

Denitrification ultimately results in a net loss of the element from the biosphere. Temporarily, translocation of the element also takes place through many other agencies like leaching of nitrate from the soil, which then finds it way to the sea or lakes, erosion removes nitrogen from the land, agricultural crops are harvesteal and removed from forms, sewage finally dumped into water bodies etc.

However, such tranlocation simply diverts the element from one system to another and it is still preserved in the biosphere. Deposition of nitrogen in the sediments is another means leading to the net loss of the element. Thus denitrification and sedimentation are the two major means of the loss of nitrogen from the biosphere and this loss has to be compensated. Nitrogen being an important element for life on the earth; that a steady- state level of nitrogen in the biosphere be maintained is imperative. In other words, it is necessary that losses must equal gains. Luckily this is by and large possible through the unique contribution made by certain microorganisms which assimilate N_2, a biologically inert gas, for synthesizing their own cell constituents through a process known as nitrogen fixation.

Nitrogen fixation

Nitrogen is very important for plant nutrition. Though present in enormous quantities in nature in the gaseous form it is not available to plants as such. N_2 therefore has to be converted into available forms before it is taken up by plants. The atmosphere contains about 10^{15} tonnes of nitrogen as N_2 and nearly ten times as much is dissolved in the ocean or occluded in rocks. Nitrogen is absorbed by the plants in inorganic forms and the atmospheric molecular N_2 has to be converted into inorganic forms for uptake. This process of conversion is typically known as nitrogen fixation. The agents of biological dinitrogen fixation belong exclusively to non-nucleate protista; the prokaryotes. Nitrogen is also fixed by many eubacteria and cyanobacteria. Fixation has also been reported by some species of Streptomyces.

Biological nitrogen fixation is accomplished by free living bacteria or blue green algae. This makes use of N_2 by non symbiotic means. Other groups of microorganisms are capable of nitrogen fixation by symbiotic associations which consists of a microorganism and higher plant.

Non Symbiotic nitrogen fixation

Azotobacter, Clostridium and several *blue-green algae* are the classical free-living microorganisms capable of fixing atmospheric nitrogen. In addition to the above, recent techniques have helped in identifying several other microorganisms capable of doing this job ; namely :

(a)   Heterotrophic bacteria : *Achromobacter, Aerobacter, Azotobacter, Bacillus, Polymyxa, Beijerinckia, Clostridium, Pseudomonas.*

(b)   Chemoautotrophic bacteria - *Methanobacillus omenlianskii.*

(c)   Blue green algae - *Anabaena, Anabaenopsis, Aulosira, Calothrix, Cylindrospermum, Nostoc, Tolypothrix.*

(d)   Photosynthetic bacteria - *Chlorobium, Chromatium, Rhodomicrobium, Rhodopseudomonas, Rhodospirillum.*

Azotobacter, however, has been the most intensively studied microorganism in relation to N_2 fixation. Azotobacters are mesophilic and are sensitive to acid pH values and to high phosphate concentrations. They rarely grow above 35°C. In the rhizospheres of plants the population of *Azotobacter* is very high and they can produce hormone like growth stimulants. *Azotobacter* is able to assimilate molecular nitrogen as well as bound forms of the element. The amount of nitrogen accumulated depends primarily on the properties of the particular strain of *Azotobacter*. The nitrogen fixation ability of the bacterium may vary greatly with various factors like cultivation, the composition and acidity of the nutrient medium, temperature, aeration; the presence of bound sources of nitrogen, the character of the carbon source, the presence of trace elements in the medium etc.

It has also been observed that the bacterium in symbiotic cultures with other microorganisms often assimilate molecular nitrogen more vigorously than in pure culture. The associated microorganisms probably supply biologically active substances to the bacterium. Some of the nitrogen assimilated is released by the bacterium into the surrounding medium. Normally the substances released are in the form of protein, amino acid and ammonia. Various organic compounds form the carbon source for *Azotobacter*. In soil the store of mobile organic matter is one of the chief factors which influences the development of the bacterium. Lack of organic substances is the chief limiting factor for the spread of the bacterium in soil which otherwise has a favourable pH and a sufficiently high phosphorus content. It has been observed that when energy stores in the soil decrease, the bacterium develops in the deeper layers. *Azotobacter* can utilize humus in soil only to an insignificant degree which does not ensure satisfactory development. So even in soils rich in humus the bacterium does not multiply greatly in the absence of fresh organic residues. It is also known that small doses of humus greatly stimulate the growth of *Azotobacter* and this can be explained as the effect of trace elements and colloid compounds present in the humus. In well cultivated horticultural soils *Azotobacter* is usually present in considerable numbers. The population of the bacterium is considerably affected by soil fertilizers. As a rule organic and phosphorus mineral fertilizers stimulate the multiplication of the nitrogen fixers while nitrogen mineral compounds often suppress their growth.

As stated earlier *Azotobacter* develops more briskly in roots than in soil.

Probably the root secretions and root litter provide a favourable substrate for the bacterium in the rhizosphere. According to Rovira (1965), however, *Azotobacter* does not constitute more than one per cent of the total micropopulation of the root region. The absolute numbers of the bacterium, however, depend on the plant species, the phase that its development has reached, the type of soil, its micropopulation and several ecological and geographical factors.

Normally it is not the plant alone, but a whole set of conditions including soil, climate, that determine the development of *Azotobacter* in the rhizosphere. Products of exosmosis from the root of different plants are of varied nature. Some root secretions may contain toxic substances. As indicated by Zayed (1963) root extract of *Populus tremula* and *Fagus sylvatica* suppresses the development of *Azotobacter*. Similarly root secretions of *Ambrosia elatior*, *Euphorbia corollata* and *Helianthus annus* are toxic to the bacterium. Oxygen deficiency in the root region also restricts the growth and development of *Azotobacter* (Vancura and Macura, 1961).

Nitrogen fixing powers of *Azotobacter* are restricted in the atmosphere. The raw materials are chiefly root secretions of the plant concerned. Exosmosis can hardly yield more than 5 per cent of the organic material systhesized by the plant. The organic substances released by the root pass through a biological filter (the microorganisms on the root surface) and so it is difficult to accept that the bacterium utilizes more than 10 percent of the products of exosmosis, which roughly come to more than 5 kg/ha. One gram of energy rich material enables *Azotobacter* to assimilate 10-15 mg of molecular nitrogen. Using 25 kg./ha. of the products of exosmosis the bacterium cannot assimilate more than 0.25-0.37 kg of nitrogen per hectare per year, which is equivalent to 1.3- 2.0 kg of sodium nitrate.

Bacteria of the genus *Beijerinckia* are similar to *Azotobacter* but differ in their resistance to acid and they can grow even at pH 3.0. *Beijerinckia* are strict aerobes. Bacteria of the genus *Beijerinckia* assimilate efficiently monosaccharides, disaccharides and starch as sources of carbon. It is resistant to soil drought and the range of temperature suitable for its development is somewhat narrow than that for *Azotobacter*. The bacteria of this genus are wide spread in the tropical soils.

Clostridium, another important nitrogen fixer, unlike the above described two bacterial genera, is an anaerobic form. A large number of species of this genus viz. *C. pasteurianum*, *C. butyricum*, *C. butylicum*, *C. beijerincki*, *C. multifermentant*, *C. pectinovorum*, *C. acetobutylicum*, *C. aceticum*, *C. felsineum*, *C. kluyveri*, and *C. madisoni* have been reported as nitrogen fixers. Emtsev (1960) observed that *C. pasteurianum* fixes nitrogen much more actively in the presence of *Bac. closteroids*, its constant companion. Probably the latter produces a series of biologically active compounds that improve the development of the former. *Clostridium* may utilize a variety of compounds for its carbon nutrition and they serve as a source of energy. This may be the reason for the almost universal distribution of *Clostridium*. Clostridia may utilize monosaccharides, disaccharides certain polysaccharides (Dextrin, Starch), polyhydric alcohols and other compounds.

However, the range of assimilable carbon sources differ for different species. Carbohydrates are decomposed anaerobically by Clostridia.

Clostridia are resistant to acid and alkali reactions and can develop within quite a wide pH range. The minimum pH for these microorganisms is about 4.7 and the maximum is above 8.5. The optimum pH range is within the limits of pH 5.9 to 8.3. Tolerance to a wide range of pH also accounts for its wide spread distribution in different soils, including those with a low pH. Being anaerobic they also tolerate a high degree of soil moisture. There is evidence that Clostridia develop when moisture is 60-80 percent of the total capacity. Usually, the upper layers of the soil, richer in organic substances, contain more Clostridia than deeper layers. In cultivated soils *Clostridium* can penetrate the deeper layers also. Temperature has varying effects on different species of the genus. *Clostridium* includes both mesophilic and thermophilic species. The mesophiles normally fix molecular nitrogen.

It has been reported that Clostridia multiply more intensively in the rhizosphere than in rest of the soil. Sometimes the rhizosphere contains ten to hundred times more cells than the corresponding nonrhizosphere soils. This is because of the availability of valuable food through root exosmosis to the bacterium. However, not all the plants promote the multiplication of *Clostridium* equally well. It multiplies more vigorously in the rhizosphere of leguminous plants than in that of grasses. In the rhizosphere most members of the genus are found in vegetative forms, which means that they are active.

Deeper into the soil the numbers of Clostridia decrease. The depth to which these bacteria penetrate depends on the profile of the soil and the status of the organic matter in it (Samtsevich, 1966).

Blue green algae :

Frank (1889) for the first time reported the ability of blue green algae or cyanophyceae to assimilate molecular nitrogen. His experimental results were, however, questioned because he did not have bacteria-free cultures of algae and nitrogen assimilation could have been due to associated bacteria. Similar doubts were also raised regarding the experimental findings of Beijerinck (1901) and Heinze (1906). Later on, Pring-Sheim (1914) and Maertens (1914) obtained pure cultures of algae and they succeeded in detecting nitrogen accumulation. The most authentic experiments were, however, done by Drewes (1928) who isolated pure cultures of three blue-green algae and convincingly demonstrated their ability to assimilate molecular nitrogen.

Nitrogen assimilation is quite widespread among the blue green algae (Singh, 1961; Fogg, 1962; Fogg and Stewart, 1965). The ability to utilize molecular nitrogen has been reported in over eighty species. With the only exception of *Chlorogloea* all other belong to the order Htormogonales. Some of the important nitrogen fixing species are : *Amorphonostoc, Anabaena, Anabaenopsis, Aulosira, Calothrix, Cylindrospermum, Fischerella, Gleotrichia, Haepalosiphon, Mastigocladus, Nostoc, Scytonema, Scytenematopsis, Stigonema* and *Tolypothrix. Chlorogloea fritschii* belongs to the order *Chroococcales*. The list of nitrogen fixing algae is growing very rapidly.

Blue green algae are widespread in different types of soil. Different species of algae seem to be the dominant nitrogen fixers in various environments. Singh (1961) stated that in the rice fields of India, the most active N_2 fixer is *Aulosira fertilissima*. In Japan Watanabe (1959) reported *Tolypothrix* tenuis as the major nitrogen fixer in Japanese rice fields. It has also been observed that in unirrigated soils *Anabaena* seems to play a major role.

Most algae responsible for nitrogen fixation are free living. However, many are symbiotic with plants. Many lichens develop symbiotic association with blue green algae. Algae are found in the cavities of liverworts where they are endophytes (Bond, 1963) and also in ferns and species of Cycas.

Individual species of microorganisms can many a time exert a positive influence on free living, nitrogen fixing algae. Bunt (1961, b) reported that species of *Caulobacter* living in the slime of blue green algae almost double the vigour of nitrogen assimilation by *Nostoc*. Similarly nitrogen fixing abilities of *Nostoc calcicola* are also enhanced in culture with nodule bacteria. There is evidence that blue green algae have a positive effect on various saprophytic bacteria. These algae intensify the multiplication and biochemical activity of many bacteria including *Azotobacter, Clostridium, pasteurianum* and *Rhizobium*.

Most blue green algae develop best in neutral or weakly alkaline media. Normally below pH 6.5 their growth is adversely affected. Blue green algae have a very distinctive response to oxygen and this warrants their being categorized as aerobes. The optimum temperature for blue green algae is close to 28°C-30°C. Sunlight is another very important factor for their development. Most of these microorganisms seem to be obligate phototrophs. Some species of blue green like *Tolypothrix tenuis* can live without utilizing solar energy and in the dark they become heterotrophs.

A correlation between photosynthesis and nitrogen fixation has been observed. The vigour of nitrogen assimilation is also influenced by the concentration of CO_2 in the medium.

Blue green algae play an important role in the accumulation of nitrogen in rice fields where conditions are optimum during a considerable part of the growing season (Singh, 1942; Prasad, 1949 and Venkataraman, 1961 a). Most investigators have reported that in these conditions the annual nitrogen increment due to the activity of blue green algae varies between 15 and 50 kg/ha which may go up to 80 kg/ha sometimes. There are also indications that blue green algae can accumulate very large quantities of nitrogen in the soil.

Rice plants have been treated with pure cultures of algae in laboratory and field conditions. In many cases the algae had a beneficial effect, but the increase in yield fluctuated greatly. In several cases there has been an appreciable increase in the rice harvested in the third or fourth years after the introduction of nitrogen fixing blue green algae. It has been observed that culture of *Anabaena* and *Nostoc* affects rice plants favourably, specially when lime, superphosphate or Mo were also added (Subramaniyam and Manna, 1966).

Blue green algae also prevent soil erosion. This is an additional advantage besides their role in accumulation of nitrogen in the soil. Singh (1961) noted the considerable importance of algae in the improvement of alkaline soil. When rain falls on alkaline soils, algae begin to develop. As a result of their activity the surface layer of soil is enriched with organic matter, the physical properties of soil improve, the pH is lowered and in the absorbing complex, sodium is replaced by calcium, with an increase in the store of P_2O_5 and nitrogen. Autotrophic blue green algae are the pioneer species in the reclamation of rock. Thus the group as a whole plays as an important role as pioneers in the formation of soil.

Besides the above microorganisms many other like *Spirillas* and *Vibrios* belonging to the family spirillaceae are able to assimilate molecular nitrogen. Molecular nitrogen can be utilized by species of *Desulfovibrio* as the sole source of nitrogen. In the family Azotobacteraceae besides *Azotobacter* which has been discussed above in detail, *Derxia gummosa* is also reported to fix molecular nitrogen. *D. gummosa*, however, is a slow-acting nitrogen fixer. D. indica is another nitrogen fixer.

In 1957 Metcalfe and Brown isolated *Nocardia cellulans* and *Nocardia calcarea* and reported them to be nitrogen fixers. Some species of *Actinomyces* like *A. aureofaciens, A. ruber, A. albidoflavus, A. fumosus, A. griseaus, A candidus, A. flaveolus* and *A. globisporus* are other nitrogen.

Symbiotic nitrogen fixation
(Nodule Bacteria of Leguminous Plants)

In previous pages the microorganisms which can fix molecular nitrogen as free living organisms have been discussed. In addition to them, there are other forms of microorganisms, which form symbiotic association with certain plants particularly in the root region and they use molecular nitrogen. The classical example of such a symbiotic association is noticed between leguminous plants and bacteria of the genus *Rhizobium*. Due to the bacterial infection roots of the plant get modified and swollen leading to nodule formation which is infact the seat of symbiosis. The symbiotic acquisition of nitrogen is accomplished in the root nodules.

Beijerinck in 1888 isolated a bacterium from the nodules of leguminous plants and he named it as *Bacillus radicicola*. A year later Prazmowski (1889) called it Bacterium radicicola. Almost simultaneously Frank (1889) proposed changing the generic name of nodule bacterium to *Rhizobium* which has been accepted universally since then.

Initially it was thought that one species of nodule bacterium infects all leguminous plants. Lateron this view proved to be false and now it has been clearly established that several rhizobial species exist and each is capable of infecting a particular leguminous species or group of related species. The different species and races of nodule bacterium have certain differences in morphological and physiological characters. Many investigators have noted a high degree of serological heterogeneity even among individual strains of nodule bacteria.

Individual cultures of nodule bacteria can infect only a definite group of leguminous plants. The group may be a large one or sometimes very small. The selectivity of nodule bacteria in relation to the host plant is termed as specificity and has been taken as the basis of classification.

Some cultures of nodule bacteria may infect with equal success a group of species or several different varieties of leguminous plants. For example, the nodule bacteria of pea, may infect vetch, the peavine and broad beans. Nodule bacteria of *Vigna* groups (cowpea) can infect many leguminous plants belonging to different sub-families (Norris, 1965).

Other cultures of nodule bacteria have a narrow adaptability, even to a particular variety, for example *Rhizobium* cultures form good tubercles only on certain species or varieties. The nodule bacteria of the clover group are considered to possess a high specificity, for they do not form tubercles on other plants. For this reason the clover group is called monolithic.

Nodules formed by active strains of bacteria are pink. This was observed long back in the nineteenth and early twentieth century. The exact chemical nature of the red pigment was, however, not known for a long time. Till 1939 the first attempt to establish the chemical nature of the pigment was unsuccessful. The pigment Pietz isolated was incorrectly identified as 5,6 quinone -2,3 dihydroxyindole -2 carboxylic acid - the intermediate products of oxidation of dihydroxyphenylalanine. In 1939, Kubo for the first time established the exact haeme nature of the pigment which was later confirmed by other workers (Barris and Haas, 1944; Keilin and Wang, 1945). It was Kubo who obtained a pure preparation of the pigment (leg haemoglobin) from soy nodules. Leghaemoglobin is present in the cytoplasm and vacuoles of the plant cell and is readily extracted with water. Smith and Jordan (1949) confirmed that there is no pigment in the cells of the bacteria and the outer cortical parenchyma. Smith (1949) observed that in different leguminous species the amount of leghaemoglobin varied between 1.09 to 3.25 mg per gram of nodule.

The haeme group of leghaemoglobin is similar to that of blood haemoglobin but the protein components differ. It has a more complex structure as indicated by molecular weight, amino acid composition and isoelectric point, which for leghaemoglobin is the lowest for any haemoglobin. It has also been noted that in annual legumes, at the end of the growing season when the process of nitrogen fixation is terminated, the red pigment turns green. Normally the change in colour begins at the base of the nodule and proceeds towards the apex. In perennial legumes, however, the greening of the nodules generally does not occur. The change of red into green nodules is irreversible.

Leghaemoglobin apparently catalyses nitrogen assimilation. It has been shown that the reduced form of leghaemoglobin is oxidised by nitrogen. Rhizobium bacteroids in the absence of oxygen may reduce leghaemoglobin. The latter is reduced by bacteria and oxidised during nitrogen fixation, promoting one of the intermediate reactions. Leghaemoglobin reduces hydroxylamine to ammonia. It reduces hydrazine to ammonia rather more slowly, though actively. All this evidence suggests a relation between the presence of leghaemoglobin and the process of nitrogen fixation in leguminous plants.

Formation of nodules in leguminous plants

Nodules are formed as a consequence of bacterial infection. Nodule bacteria usually penetrate the roots through the the root hairs. Occasionally they also enter through damaged epidermal cells and cortical cells. This normally happens at the sites of branching of the lateral roots. When the root of a leguminous plant develops in the soil, it accumulates an abundant microflora peculiar to the rhizosphere of the plant to which it belongs. Species of leguminous plants give rise to conditions specially favourable to the multiplication of its own nodule bacteria. According to Rovira (1957) the stimulating effect of roots affects the nodule bacteria even at a distance of 20-30 mm. A large number of nodule bacteria in the rhizosphere are necessary to ensure completion of the process of infection.

From the root region of the plants as indicated earlier, a large number of substances are released in the form of root leakage. Sugars, organic acids, amino acids, nucleotides, vitamins, enzymes are some such substances. Under the influence of nodule bacteria tryptophan is converted to indole-3- acetic acid (IAA) which causes a distinctive change in the shape of root hairs, bending them to the shape of an umbrella handle. Nodule bacteria also form an extracellular mucilage of a polysaccharide nature - and this steps up the plants' production of the enzyme polygalacturonase, which is thought to act on the wall of the root hair, making it more plastic and permeable to bacteria.

Penetration of root by bacteria

Nodule bacteria penetrate the root hair at certain places where the wall is sufficiently permeable. The wall of the root hair contains cellulose and pectin substances that nodule bacteria cannot decompose. There are various views regarding the exact mode of entry into the root. According to Nutman (1956) bacteria push through the tip of the root hair when it is covered only by the first wall layer of pectic substances, polygalacturonides, galoctanases, arabines, hemicellulases, waxes and polypeptides. The fibres of cellulose comprise most of the wall, in the form of a matrix. The gaps between fibres are upto 0.3 m wide and are not filled with calcium pectate. Therefore with some stretching of the membrane the bacteria can pass into the hair.

Rudakov and Birkel (1954) were of the opinion that soil bacteria destroy pectin of the cell wall and thereby making it accesible to nodule bacteria.

Ljunggren and Fahraeus (1959, 1961) opined that enzyme polygalacturonase is

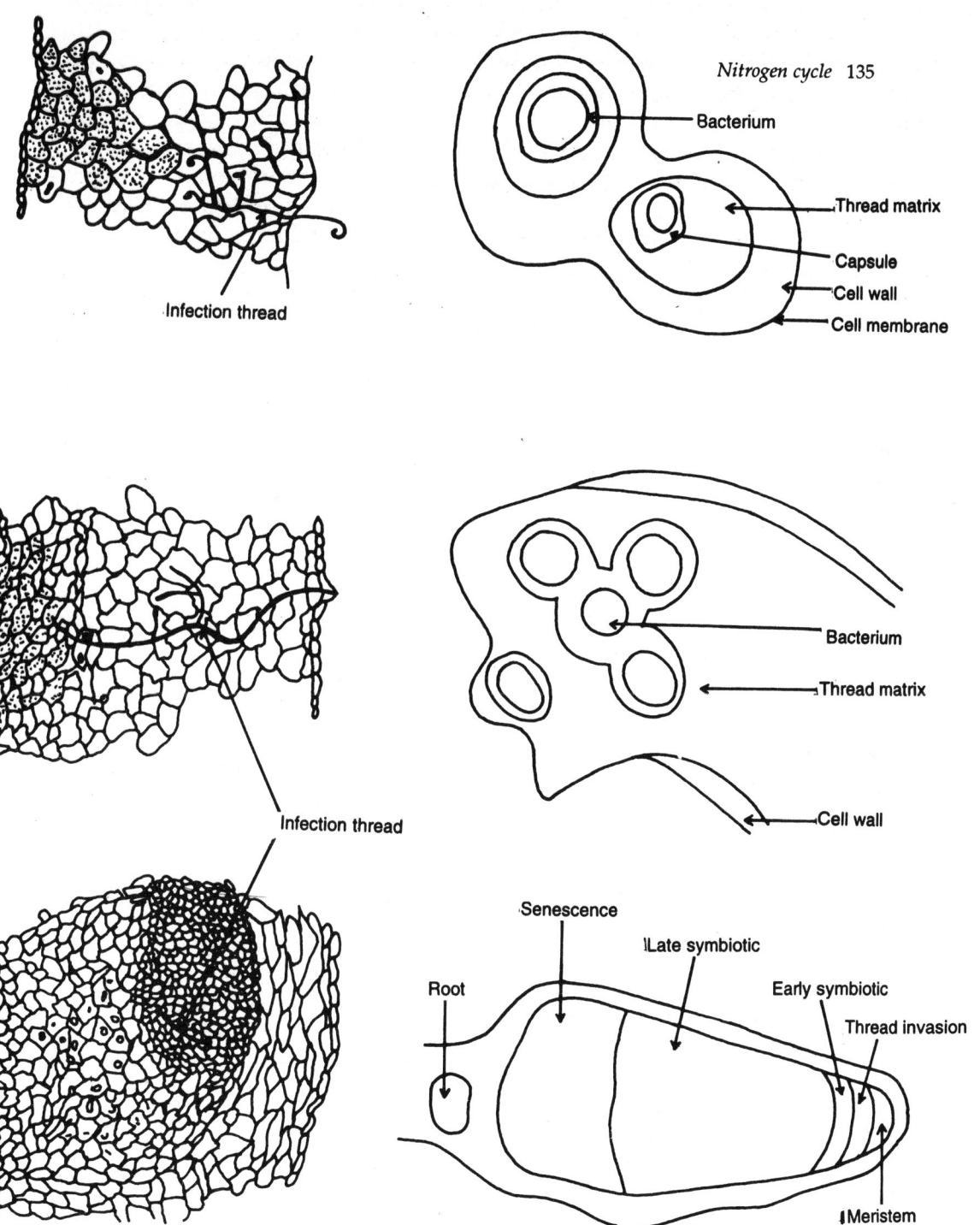

Nitrogen cycle 135

Fig. 6.1(c) Stages in the infection by Rhizobia

essential for increasing the permeability of the wall of the root hair. It is always present in small amounts in the root hair and evidently, by causing partial dissolution of the components of the wall, it enables the cell to stretch. Substances akin to polysaccharides, secreted by the nodule bacteria, greatly promote the formation of polygalacturonase. In turn, this leads to softening of the root hair wall which enables the nodule bacteria to enter more easily.

With in hair, the nodule bacteria form a so called 'infection thread' - a hypha-like mucilaginous mass in which are buried the multiplying rod-shaped bacteria. The bacteria in the thread have a cytoplasm membrane and a cell wall 8 nm thick. On the outside the bacteria are surrounded by a microcapsule. The infection thread moves towards the base of the hair and the cells of the epidermis. This distance, about 100-200 μ, is covered in 1-2 days at about 5-8 μ per hour. The thread may move as a result of the pressure which builds up when the bacteria develop within it. As a rule one thread is formed for each root hair, although there are cases when several threads are seen in a single hair or several root hairs have a common thread.

Membrane is formed around the infection thread which is supposed to be as a defence reaction of the cytoplasm of the host cell. The growing apex of the thread, however, is free of membranous covering. On coming into contact with the host cell wall of the infection thread, the wall usually thickens. The infection thread then crosses the intercellular spaces and by this time the nearest plant cell proliferates and forms a funnel around the thread. Lengthening of this funnel produces a tubular sheath surrounding the infection thread.

It has been reported that nodule bacteria multiply in tetraploid cells and when the infection thread reaches such cells of the cortex, some nodule bacteria pass from the thread into the cytoplasm of these cells and there begin to multiply. As soon as the cytoplasm of the plant cell is infected by nodule bacteria, division is induced in the cell concerned and also in the adjacent uninfected cells. Division usually begins one to two cell layers from the end of the infection thread. As a result of this, the process of nodule formation begins.

Thereafter the infection thread branches and spreads further through the tetraploid cells. The cortex and conducting vessels of the nodule are formed from the diploid tissue of the plant. The bacteria then pass out of the infection thread, specially from its apex if this is not covered with a membrane or after rupture of the vesicle like evaginations formed by the infection thread in the plant cells. These evaginations are usually devoid of cellulose sheath and therefore the bacterial cells surrounded, as it were, by vacuole cavities, readily branch off into the cytoplasm of the host cells.

Bacteria which have passed into the cytoplasm elongate to rods and they continue to multiply. Later these cells are transformed into bacteroids. The latter cannot divide but greatly increase in size. The bacteroids gradually swell and begin to occupy much of the plant cell, which usually expands considerably. As a rule the formation of bacteroids is associated with the appearance of leg-haemoglobin.

While the bacteroids are in the process of formation, the mitochondria and cell plastids move towards the cell membrane and line up along it. At this time only leghaemoglobin is observed in the nodules which is involved in the process of nitrogen fixation. In the apical region of the nodules, bacteroids are normally absent and the maximum concentration is noted in the central region. In the beginning the nodules are whitish in colour becoming pink at the time of optimum activity. When the active life of the nodule comes to an end the bacteriods under go lysis and this usually happens at the flowering stage of the plants. The necrosis begins at the centre of the bacteriod region or close to the base of the nodule and progresses to the periphery. It has been observed that nodules are short lived in annual plants, however, in perennials they may survive and are active for several years. By the end of the season the bacteriod tissue degenerates but the whole nodule doesnot die off, and in the following year it retains its activity and function again.

Biochemistry of N2 fixation

Atmospheric nitrogen is fixed in nodules. There have been many hypotheses concerning the chemistry of nitrogen fixation. The reduction of nitrogen to ammonia is catalysed by an enzyme complex known as nitrogenase. Electrons are passed to the Fe protein part of the nitrogenase complex. The latter is also known as dinitrogenase reductase. The reduced Fe protein joined to two Mg ATP molecules, inturn transfers electrons singly to the Fe-Mo-protein part of the complex, releasing MgADP and Pi. To a site associated with Molybdenum, electrons flow from the reduced Fe-Mo-protein to various substrates, of which the natural ones are N_2 and H^+. Many other substrates can also be reduced. The most widely known is acetylene which diverts all of the available electrons towards production of ethylene. In natural conditions, i.e. in air, electrons are divided between the reduction of nitrogen and protons, giving ammonia and hydrogen gas.

Becker and Evans (1980) purified a ferredoxin from *Rhizobium japonicum* (which nodulate soy bean) and showed that it could reduce the Fe protein. How ferrodoxin is subsequently reduced is not yet clear. One suggestion is that an intact bacterial membrane could help the flow of electrons to ferredoxin, if either (a) the membrane potential is sufficiently large or (b) there is a proton motive force sufficient to drive reversed electron transport. A combination of (a) and (b) is also possible.

It is worth noting that many of the organisms capable of nitrogen fixation also have an enzyme that reduces protons to molecular hydrogen using ferredoxin as the reductant.

$$2H^+ + \text{ferredoxin (red)} \longrightarrow H_2 + \text{ferredoxin (oxid)}.$$

This enzyme, called hydrogenase, produces hydrogen.

Under some circumstances (such as the absence of N_2) the enzyme nitrogenase can reduce protons to evolve hydrogen. This reaction unlike the hydrogenase reaction is dependent on ATP energy.

The nitrogenase reaction, reducing N_2 to ammonia, has a very high energy requirement. It appears that 12 ATP are required for the reduction. Many of the nitrogen fixing organisms are anaerobic or only fix nitrogen under anaerobic conditions. ATP is evidently supplied by anaerobic respiration.

It would seem logical to conclude that symbiosis creates a situation in which energy is supplied by the host, the legume plant in the case of *Rhizobium*. In the legume nodule there is a protein comparable to myoglobin, called leghaemoglobin. The leghemoglobin has a higher affinity for oxygen than myoglobin and hemoglobin. The primary role of leghaemoglobin in nitrogen fixation is assumed to be oxygen binding in the vicinity of the oxygen sensitive enzyme nitrogenase. Most of the leg hemoglobin is located in the plant cell cytoplasm but some, possibly of considerable physiological significance, bathes the bacteroids within their envelopes.

There is now overwhelming evidence that leghaemoglobin acts, as does haemoglobin in blood, as an oxygen - carrying pigment. One of the nodules chief dilemmas is to allow sufficient oxygen to reach the bacteroids for ATP synthesis without causing inactivation of nitrogenase. What is needed is a high flux of oxygen at a low concentration. The oxygen - binding properties of leghaemoglobin achieve this.

There are several distinct protein moieties in leghaemoglobin from the same and from different species; the proteins differ in amino-acid sequence and oxygen binding properties and each is coded by a different host gene. Uheda and Syono (1982) observed that within individual nodules, the ratio of different components varies with age and they suggested that this may result in an improvement in oxygen carrying capacity as nodules age.

Leghaemoglobin is only the most obvious of a number of nodule- specific molecules. Host genes are known to code for 18-20 proteins tentatively termed 'nodulins', which are only produced in effective nodules (Auger and Verma, 1981).

Ammonia - assimilating path ways

It is general dogma that ammonia must be assimilated quickly. In nitrogen fixing system there are two reasons for this, the first is that ammonia is toxic and the second that ammonia represses nitrogenase synthesis. Dilworth and Glenn (1982) observed that rhizobia can metabolize high concentrations of ammonia, however, their host cells probably cannot. The synthesis of nitrogenase in rhizobia may be rather more resistant to ammonia than in other organisms and in any case, atleast in intermediate nodules, there may be a little concurrent co-located nitrogenase synthesis and nitrogenase activity. There is considerable evidence that the major path way of ammonia assimilation is into the amide group of glutamine using the high affinity glutamine synthetase rather than the lower affinity glutamine dehydrogenase system. The enzymes are located mainly in the host cells. In other words, once the bacteroids have reduced the nitrogen gas, the ammonia is handed on to the host cell for assimilation - a good division of labour

between the symbionts (Flow chart I).

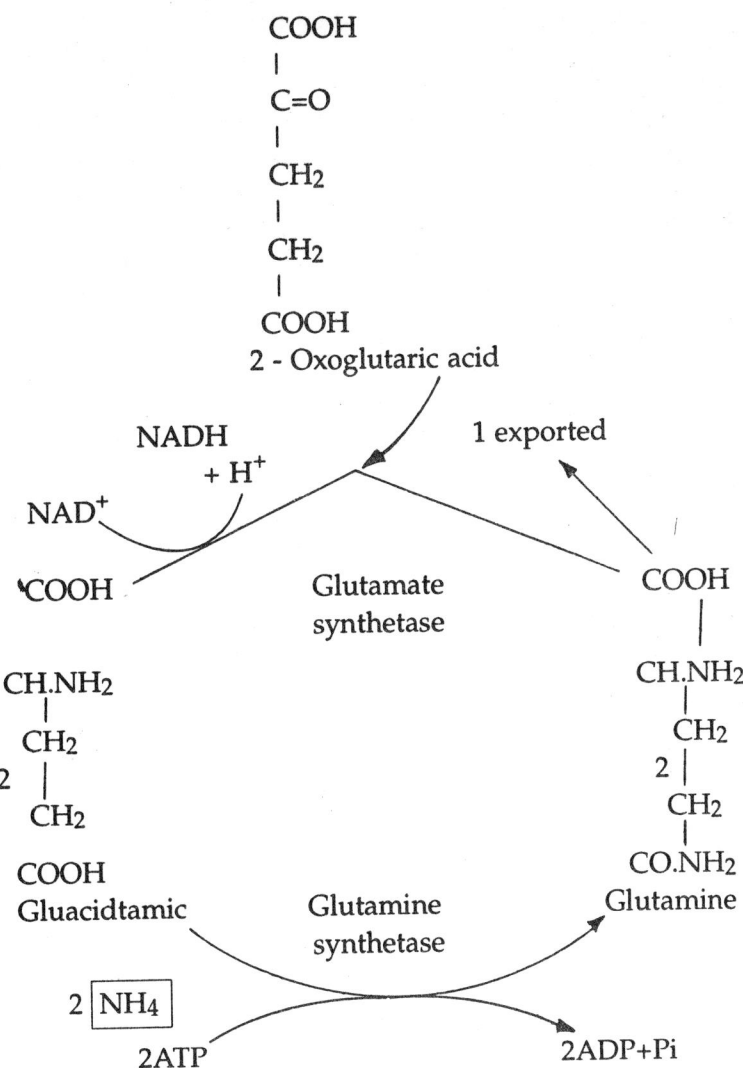

Flow chart I. Incorporation of ammonia into glutamine as in legume nodule host cell cytoplasm.

Amide synthesis

Temperate legumes export from their nodules principally the amide glutamine and asparagine. The normal assimilatory pathway couples glutamine synthetase with glutamate synthase, as outlined in flowchart I. The bold arrows indicate how net glutamine synthesis may occur at the expense of one molecule each of 2 - oxoglutarate and reduced pyridine nucleotide and two of ATP. $NADH + H^+$ may be regarded as 3ATP equivalents, since it might otherwise be used in oxidative phosphorylation.

Ureide synthesis

Recently it has been found that a number of legumes of tropical and sub-tropical origin, such as soybean, Phaseolus and Vigna, export the ureides allanoin and allantoic acid. Current evidence favours the following pathways for ureide synthesis in host cells (Flow chart 3). The initial assimilation of ammonia is into glutamine and glutamate, (Flow chart 1) and then transamination reaction yields directly or indirectly the other purine precursors, glycine and asparate. The process of purine synthesis appears to be broadly similar to that in other organisms. The likely origin of the various atoms of the heterocyclic ring are indicated in Flow chart II (A & B) and III.

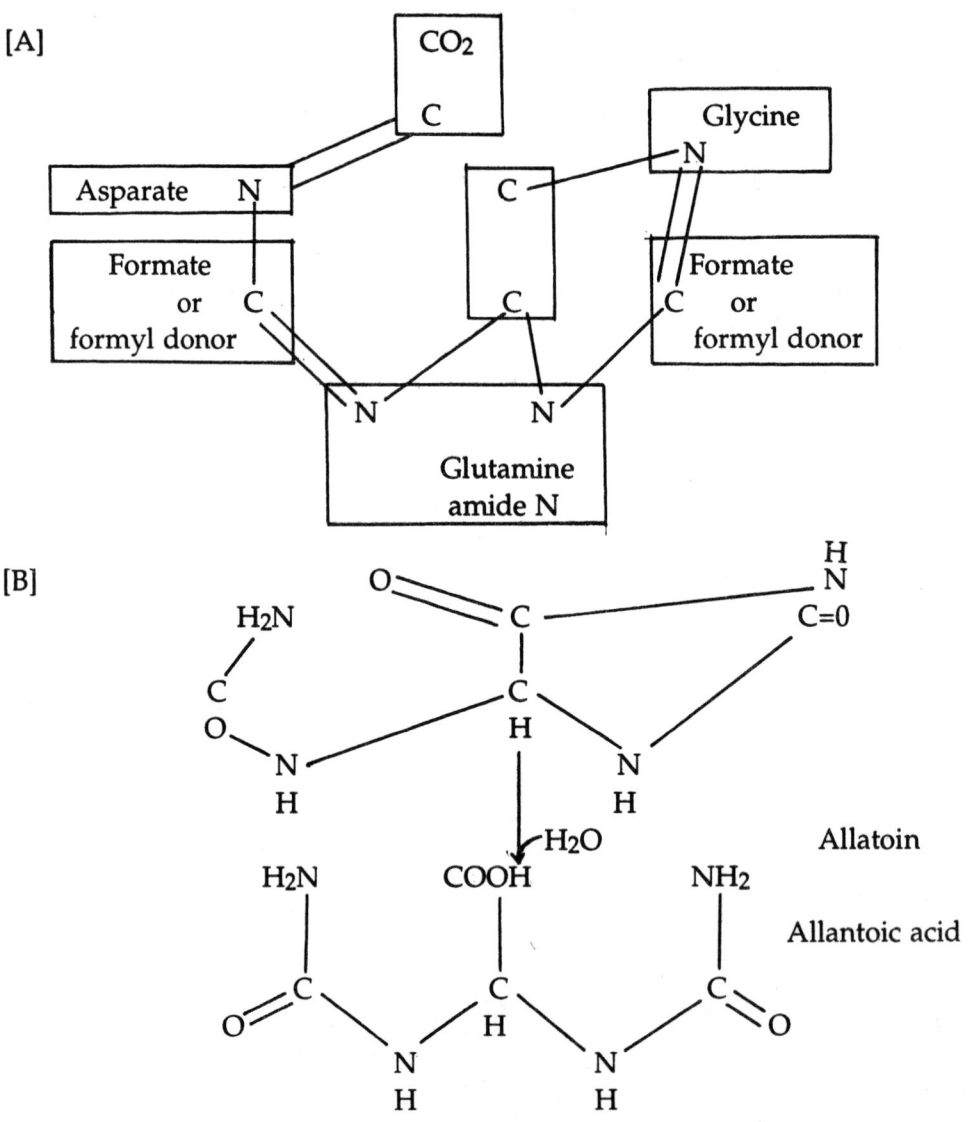

Flow chart II. Ureide synthesis

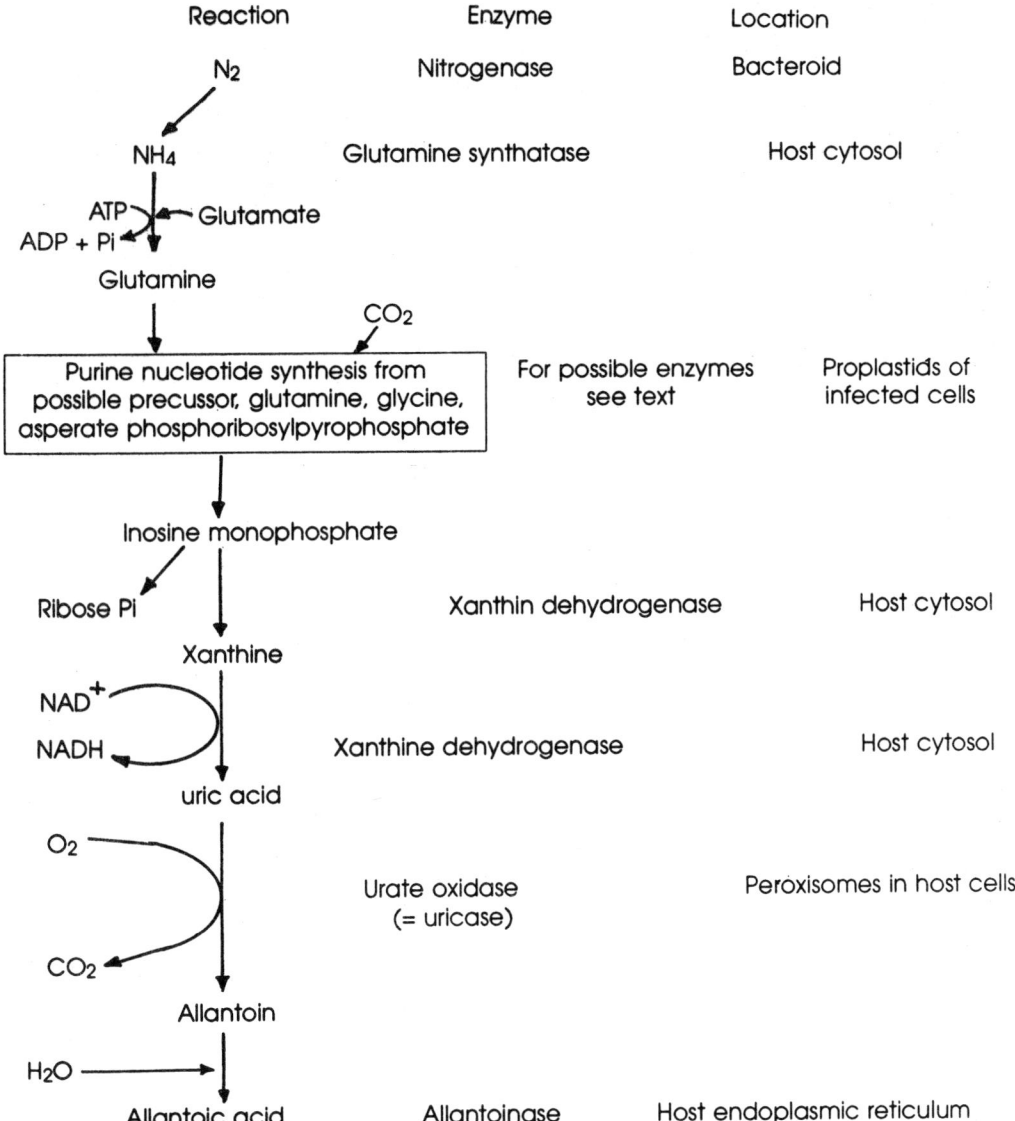

Reaction   Enzyme   Location

N_2 Nitrogenase   Bacteroid

NH_4 Glutamine synthatase   Host cytosol

ATP ← Glutamate
ADP + Pi
Glutamine

CO_2

Purine nucleotide synthesis from possible precussor, glutamine, glycine, asperate phosphoribosylpyrophosphate   For possible enzymes see text   Proplastids of infected cells

Inosine monophosphate

Ribose Pi

Xanthine   Xanthin dehydrogenase   Host cytosol

NAD^+
NADH   Xanthine dehydrogenase   Host cytosol

uric acid

O_2 Urate oxidase (= uricase)   Peroxisomes in host cells

CO_2

Allantoin

H_2O

Allantoic acid   Allantoinase   Host endoplasmic reticulum

Flow chart III. Probable pathway of ureide biogenesis in nodules of certain legumes, generally Of tropical/subtropical origin.

Whereas the amounts of ATP energy or its equivalent used in synthesizing amides are well established, there are several incompletely understood reactions in purine synthesis which may affect the total amount of ATP used. The reactions upto the formation of free purine, xanthine, are necessary for nodule growth as well as to produce export products. Subsequent reactions are specific to the formation of export products. This is indicated by the high correlation between the activities of the relevant enzymes and nitrogenase (Atkins *et. al.* 1982; Reynold *et. al.* 1982). NADH is produced during xanthine oxidation to uric acid (Flow

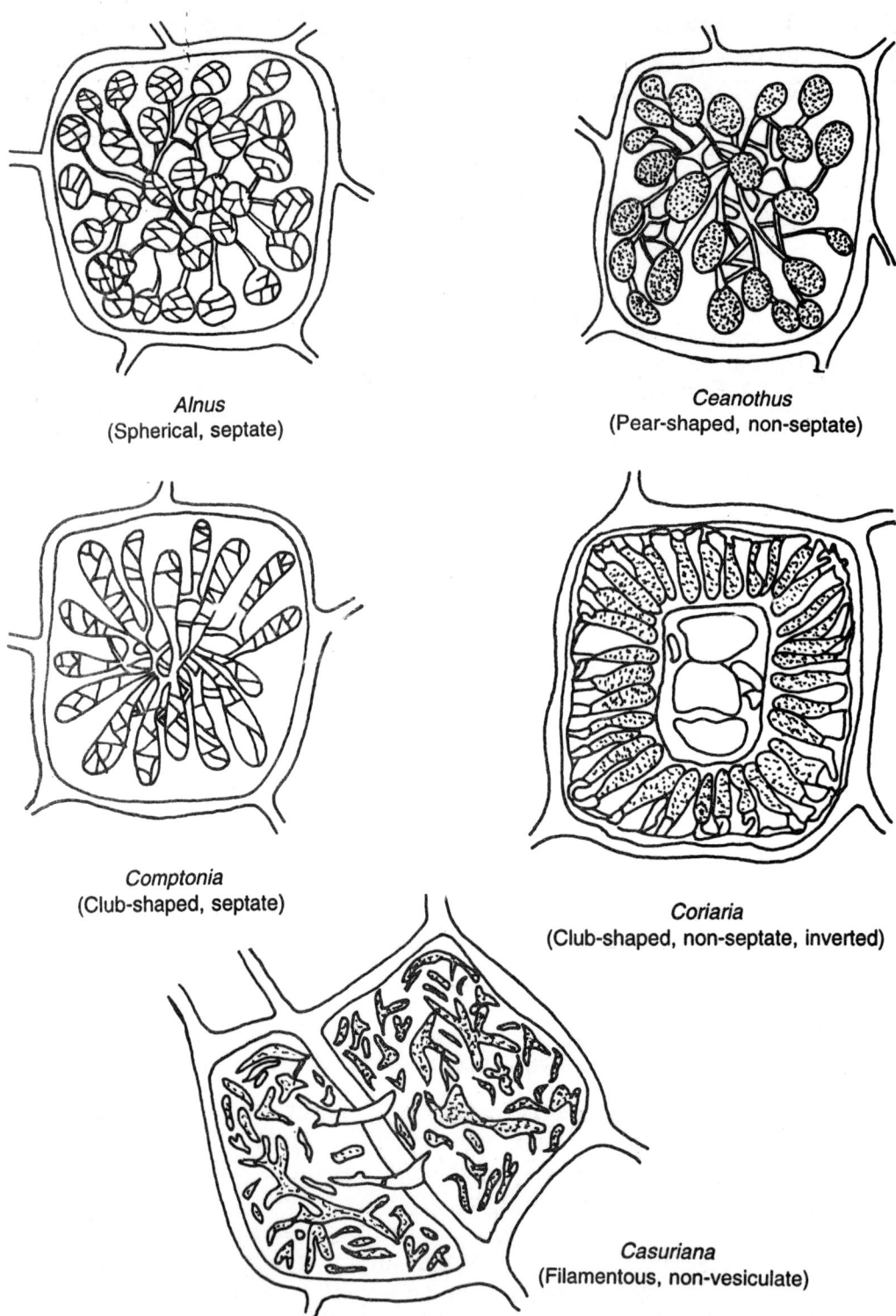

Alnus
(Spherical, septate)

Ceanothus
(Pear-shaped, non-septate)

Comptonia
(Club-shaped, septate)

Coriaria
(Club-shaped, non-septate, inverted)

Casuriana
(Filamentous, non-vesiculate)

Fig. 6.1(d) Frankia infected root nodule cells of different hosts

chart 3). The latter is then converted to allantoin. If the nodules are short of oxygen, ureide synthesis may be curtailed. Even otherwise oxygen is usually the major factor affecting ureide production.

Sprent (1980) observed that both allantoin and allantoic acid are relatively insoluble compounds and under cool conditions this low solubility could restrict export from nodules.

Because of the importance of nitrogen to agriculture, there is much interest in the genetics of nitrogen fixation. Experiments of Streicher and Valentine (1973) showed that it was possible to transfer DNA with nitrogen fixation genes (called nif for nitrogen fixation) from a nif$^+$ (nif - positive) strain of Klebsiella to a nif strain. Because of our knowledge of molecular biology and the capacity of genetic transfer by the techniques of transduction and conjugation there is now the possibility of introducing nif genes into higher plants. Of course even if the nif$^+$ genes were introduced into the plant gernome, the problem of regulation would still remain. Nevertheless, when it is realised that legumes can be grown without the addition of nitrogen fertilizer, the importance of plants that can fix nitrogen is apparent.

Little is clearly known about the regulation of nitrogenase, but ammonia represses its synthesis. The ammonia effect, however, seems to be indirect and operates through the enzyme glutamine synthetase. Glutamine synthetase produces glutamine from ammonia and glutamate. Apparently ammonia converts glutamine synthetase to a form that acts on the nif genes, preventing transcription and the ultimate synthesis of nitrogenase.

Once ammonia or another fixed form of nitrogen is taken up by the plant, it is converted to organic nitrogen. The assimilation of nitrate and ammonia is discussed earlier.

Actinorrhizal systems

Besides legumes some more plants have nodules. This fact was known since the beginning of the century when Wobbe, Schmid, Hiltner and Hotter in 1892 reported this fact. Largely through the efforts of Bond their importance became recognized in the 1960s and 1970s. A prime example of this group is alder (*Alnus*) which forms nodules. The latter do not contain rhizobia, but they are diazotrophic. Other plants which form comparable nodules are generally shrubs or trees, for example : *Shepherdia* and *Ceanothus* (order Rhamnales), *Myrica* and *Comptonia* (order Myricales), *Casuarina* (order Casuarinales), *Coriaria* (order Coriariales), *Dryas* (order Rosales) and many others. Over 200 species of diazotrophic nodulating non-legumes are now known. Though effective nodules of these plants are often pink, this colour is due to anthocyanins, not leghaemoglobin. The symbiotic microorganisms responsible for nodulation and diazotrophy proved elusive for many years and it was only in the late 1970-s that they were unequivocally identified and characterized. Becking (1970) assigned these endophytes to the generic name *Frankia* which is included in the sub-family of the *Streptomyces* called Frankiaceae.

The associations formed by *Frankia* have been given the name actinorrhizal association. The process of infection is almost similar to the legumes (Newcomb, Callaham, Torrey and Peterson, 1979). Root hair deformation usually occurs and infection threads have been reported in *Comptonia* and some other plants. Infection apparently leads to penetration of the cell cytoplasm by microbial filaments which displace the nucleus. Effective diazotrophy is associated with the appearance of vesicle - like bodies which appear to be the loci of nitrogenase activity (Van straten, Akkermans and Roelofsen, 1977).

Tjepkema *et. al.* (1980) reported that *Frankia* species isolated from Comptonia *peregrina* is capable of nitrogen fixation. Fixation in axenic culture is associated with the appearance of vesicle - like bodies resembling those seen in the Comptonia nodule; the process seems to be relatively insensitive to oxygen.

Leaf nodule associations

Some tropical plants such as *Psychotria* have tiny nodules in their leaves within which bacteria can grow. Centifanto and Silver (1964) isolated a diazotrophic *Klebsiella*, which they called *K. rubiacearum*, from such nodules and some suggestions arose that nitrogen might be fixed in the leaf nodules. However, there is little doubt that phyllosphere is a good habitat for microbes, including diazotrophs. Azotobacters and Beijerinekia (Ruinen, 1956) may inhabit the phyllosphere of plants, particularly tropical plants such as coffee, cotton and sugar cane. Bessems (1973) presented evidence that Klebsiellae fix nitrogen in the phyllosphere of temperate maize as well as tropical grass. The diazotrophs supply fixed nitrogen to the soil as they are washed off by rains but it is difficult to find satisfactory quantitative data for this nitrogen input.

Azospirillum species - the grasses and the cereals

In 1976 Dobereiner and Day reported a new diazotrophic association between a curved bacterium and the grass *Digitaria decumbens*. Azospirilla are widely distributed in tropical and temperate soils (Tyler *et. al.* 1979). Azospirilla are interesting because of their broad host range (Day and Dobereiner, 1976) and in particular, their ability to colonize the roots of agricultural cereals such as maize and sorghum. It has now been convincingly proved that azospirilla do colonize the rhizosphere of grasses and cereals such as maize and they do penetrate the inner root tissues and that they continue to fix nitrogen in that environment.

Factors affecting nitrogen fixation

Various environmental factors like temperature, water stress, water logging, salinity and combined nitrogen affect nitrogen fixation. These factors also affect the nodule growth, often differentially affecting the process of cell division and enlargement. Extremes of soil pH and supplies of essential nutrients such as P, K, Mo and Fe affect both nodule development and physiology. In all the species studied, nodules with intermediate growth have better recovery potential than those with determinate growth because they can resume merisatemstic activity and hence make new nitrogen fixing tissue.

Temperature

Generally for any one host - rhizobial combination, maximum rates of activity are obtained over a temperature range of 10°C. The actual values vary widely, usually around 15-25°C for temperate and 25-30°C for tropical species. Nitrogen fixation and/or assimilation, gaseous diffusion and solubility and respiratory activity are affected by temperature.

Water stress

The effects of water stress like temperature are complex. Under conditions of photosynthetic limitation, nitrogen fixation may fall rapidly following stomatal closure. However, by maintaining a water supply to nodulated roots at the same time as exposing the shoot system to a very low relative humidity, it has been possible to reduce photosynthesis in soybean to a very low level without affecting nodule activity. Under field conditions water stress may seriously reduce nitrogen fixation.

Water logging

Intermittent water-logging reduces the oxygen supply to the nodules and is therefore inhibitory to the nitrogen fixation. Prolonged exposure to wet soils may result in the formation of nodules with more air spaces and thus improved tolerance to waterlogging.

Salinity

Generally, nodulated crop plants do not like saline conditions, although at low salt levels (25 mol m-3 NaCl) they may show compensation by producing larger nodules (Youself and Sprent, 1983). Wilson (1970) observed that different legumes responsd differently to salinity. He further observed that plants grown on combined nitrogen are more tolerant than those dependent on fixed nitrogen.

Combined nitrogen

The reasons why combined nitrogen reduces nodulation as well as activity of performed nodules have been studied extensively. There have been two schools of thought

(a) that photosynthate is diverted away from nodules towards areas where combined nitrogen is being assimilated, and

(b) that in the case of nitrate, bacteroid nitrate reductase produces nitrite in concentrations high enough to inhibit nitrogenase activity. Streeter (1983) suggested that nitrate reduction in the host cells of nodules may be the controlling reaction.

Reference[*]

Abbott, L.K. and Robson, A.D. (1985). Formation of external hyphae in soil and four species of vesicular arbuscular mycorrhizal fungi. New Phytol., **99**, 245-255.

Acharya, C.N. (1935). Biochem. J., **29**, 1116-1120.

Ajello, L. (1956). Science, **123**, 876-879.

Alexander, F.E.S. and Jackson, R.M. (1954). Nature, Lond. **174**, 750-751.

Alexander, M. (1961). Introduction to Soil Microbiology. John Wiley and Sons, Inc. New York and Lond.

Alexander, C.; Alexander, I. J. and Hadley, G. (1984). Phosphate uptake by *Goodyera repens* in relation to mycorrhizal infection. New Phytol. **97**, 401-411.

Allen, M.F. (1982). Influence of vesicular arbuscular mycorrhizae on water movement through *Bouteloua gracilis* (H.B.K.) Lag ex. Steud. New Phytol. **91**, 191-196.

Bagyaraj, D.J. (1984). Biological interaction with VA mycorrhizal fungi. In: "VA Mycorrhiza" (Powell, C.LI. and Bagyaraj, D.J. eds.) CRC Press. Boca Raton, FL, pp. 131-153.

Baltruschat, H. and Schonbeck, F. (1975). The influence of endotrophic mycorrhiza on the infestation of tobacco by *Thielaviopsis basicola*. Phytopathol. Z., **84**, 172-188.

Band, R.N. (1959). J. Gen. Microbiol. **21**, 80-95.

Barea, J.M.; Azcon-Aguilar, C. and Azcon, R. (1987). Vesicular- arbuscular mycorrhiza improve both symbiotic N_2 fixation and N- uptake from soil as assessed with a N^{15} technique under field conditions. New Phytol. **106**, 717-725.

Barton, R. (1957). Nature, Lond. **180**, 613.

Bartlett, E.M. and Lewis, D.H. (1973). Surface phosphate activity of mycorrhizal roots of beech. Soil Biol. Biochem. **5**, 249-257.

Beckwith, T.D. (1911). Phytopathology, **1**, 169-176.

Bethlenfalvay, G.J., Pacovsky, R.S., Brown, M.S. and Fuller, G. (1982 b). Mycotrophic growth and mutualiastic development of host plants and fungal endophyte in an endomycorrhizal symbiosis. Plant and Soil, **68**, 43-54.

Bowen, G.D. and Rovira, A.D. (1961). Plant and Soil, **15**, 166-188.

Bowen, G.D. (1973). Mineral nutrition of ectomycorrhizae. In: "Ectomycorrhizas : Their ecology and physiology" (Marks, G.C. and Koslowski, T.T. eds.). Academic Press, New York, pp. 151-205.

---

[*]Only selected references cited.

Bray, J.R. and Gorham, E. (1964). Adv. Ecol. Res. **2**, 101-152.

Breed, R.S.; Murrary, E.G.D. and Smith, N.R. (1957).

Bergey's Manual of Det. Bact." 7th Ed. Williams and Wilkins, Baltimore.

Bromfield, S.M. and Skerman, U.B. (1950). Soil Sci. **69**, 337- 348.

Brown, JC. (1958). Trans. Br. Mycol. Soc. **41**, 81-88.

Burbank, W.D. (1942). Physiol. Zool. **15**, 342-362.

Burges, N. A. and Fenton, E. (1953). Trans. Br. Mycol. Soc. **36**, 104.

Burges, A. (1958). The Microorganisms in the soil, Hutchinson, London.

Burges, A.; Hurst, H.M. and Walkden, B. (1964). Geochim. Cosmochim. Acta **28**, 1547-1554.

Burges, A. and Raw, F. (1967). "Soil Biology", Acad. Press, London and New York.

Buxton, E.W. (1962). Ann. Appl. Biol. **50**, 269-282.

Bywater J. and Hickman, C.J (1959). Trans. Br. Mycol. Soc. **42**, 513-524.

Calder, E.A. (1957). J. Soil Sci., **8**,, 60-72.

Chaboud, A. (1983). Isolation, purification and chemical composition of maize root cap slime. Plant and Soil, **73**, 395- 402.

Chakravarty, P. and Unestam, (1987). Mycorrhizal fungi prevent disease in stressed pine seedlings. J. Phytopath., **118**, 335-340.

Chesters, C.G.C. (1948). Trans. Br. Mycol. Soc., **30**, 100-117.

Chesters, C.G.C.; Apinis, A. and Turner, M. (1956). Proc. Linn. Soc. Lond. **166**, 87-97.

Chesters, C.G.C. and Parkinson, D. (1959). Plant and Soil, **11**,145- 156.

Cholodny, N. (1930). Arch. Mikrobiol. **1**, 620-652.

Clark, F.E. (1940). Trans. Kans. Acad. Sci., **43**, 75-84.

Clark, F.E. (1948). Proc. Soil Sci. Soc. Am. (1947), **12**, 239- 242.

Clark, F.E. (1949). Adv. Agron. **1**, 241-288.

Clarkson, D. (1985). Factors affecting mineral nutrition acquisition by plants. Annu. Rev. Plant Physiol., **36**, 77-115.

Conn, H.G. (1918). Tech. Bull. N.Y. St. Agric. Exp. Stn., **64**, 1- 20.

Conn, H.G. (1928). Soil. Sci., **25**, 263-272.

Cooke, W.B. (1958). Bot. Rev., **24**, 341-429.

Corbaz, R.; Gregory, P.H. and Lacey, M.E. (1963). J. Gen. Microbiol., **32**, 449-456.

Cox, G. and P.B. Tinker, (1976). Translocation and transfer of nutrients in vesicular-arbuscular mycorrhizas. I. The arbuscule and phophorus transer : A quantitative ultrastructural study. New Phytol., **77**, 371-378.

Crawford, D.V. (1956). Rapports, 6th Intl. Cong. Soil Sci., Paris, C, 197-202.

Cromer, D.A.N. (1938). The significance of the mycorrhiza of *Pinus radiata*. Bull. For. Bur.

Aust., **16**, 1-19.

Croney, D. (1952). Geotechnique, **3**, 1.

Cummins, C.S. and Harris, H. (1958). J. Gen. Microbiol., **18**, 173-189.

Dale, E. (1912). Ann. Mycol., **10**, 452-477.

Dale, E. (1914). Ann. Mycol., **12**, 33-62.

Davis, R.M. (1980). Influence of *Glomus fasciculatus* on *Thielaviopsis basicola* root rot of citrus. Plant Dis., **64**, 839- 840.

Davis, R.M. and Menge, J.A. (1980). Influence of Glomus fasciculatus on Phytophthora root rot of citrus. Phytopathology, **70**, 447-452.

De Rose, H.R. and Newman, A.S. (1947). Soil Sci. Soc. Am., Proc. **12**, 222-226.

Dobbs, C.G. and Huison, W.H. (1953). Nature, London, **172**, 197- 199.

Dodd, J.C. Burton, C.C. Burns, R.G. and Jeffries, P. (1987). Phosphate activity associated with the roots and the rhizosphere of plants infected with vesicular arbuscular mycorrhizal fungi. New Phytol., **107**, 163-172.

Dowding, E.W. (1959). Ecology of *Endogone*. Trans. Brit. Mycol. Soc., **42**, 449-457.

Drechster, C. (1941). Biol. Rev. **21**, 377-439.

Fedorov, M.V. (1952). "Biological fixation of atmospheric nitrogen". 2nd ed. Gosudarstv. Izdatel. Sel'skokhoz. Let. Moscow (Russian).

Fay, P. and Fogg, G.E. (1962). Arch. Mikrobiol., **42**, 310-321.

Feher, D. (1948). Erdesz. Kiserl., **48**, 57-93.

Fletcher, J.E. and Martin, W.P. (1948). Ecology, **29**, 95-100.

Flint, E.A. (1958). N.Z. J. Agric. Res. **1**, 991-997.

Fogel, R. and Hunt, G. (1979). Fungal and arboreal biomass in a western Oregon Douglas-fir ecosystem : distribution patterns and turnover. Can. J. For. Res., **9**, 245-256.

Francke, H.L. (1934). Beitrage zut Kentius der Mykorrhiza von *Monotropa hypopitys* L. Analyse and synthese der symbiose. Flora (Jena), **129**, 1-52.

Frank, A.B. (1885). Uber die anf Wurzely mliose berunhende Ernabrung gewisser Baume durch unterirdische. Pilze. Ber. dt. bot. Ges. **3**, 128-145.

Frankland, J.C. (1966). J. Ecol., **54**, 41-64.

Frenzel, B. (1960). Planta, **55**, 169-207.

Garrett, S.D. (1950). Ecology of root inhabiting fungi. Biol. Rev., **25**, 220-254.

Garrett, S.D. (1956). "Biology of root infecting Fungi". Cambridge University Press, London.

Gerdemann, J.W. and Nicolson, T.H. (1963). Trans. Brit. Mycol. Soc., **46**, 235-244.

Gilman, J.C. (1957). "A Manual of Soil Fungi". Jowa State College Press, Ames, Jowa.

Goodey, T. (1915). Ann. Appl. Biol., **1**, 395-399.

Graf, G. (1950). Zentd. Bakt. Parasitkde Abt. II, **82**, 44-69.

Graf, W. (1958). Arch. Hyg. Bakt. **142**, 267-275.

Graham, J.H. and Syvertsen, J.P. (1984). Influence of vesicular arbuscular mycorrhiza on the hydraulic conductivity of roots of two citrus root stocks. New Phytol., **97**, 227-284.

Graustein, W.C. Cromachk, K. and Sollins, P. (1977). Calcium oxalate occurrence in soils and effect on nutrient and geochemcial cycles. Science, **198**, 1252-1254.

Grein, A. and Meyers, S.P. (1958). J. Bact., **76**, 457-463.

Groscop, J.A. and Brent, M.M. (1964). Can. J. Microbiol., **10**, 579-584.

Haarlov, N. and Weis-Fogh, T. (1953). Oikos, **4**, 44-57.

Haarlov, N. and Weis-Fogh, T. (1955). In "Soil Zoology" (D.K. McE. Kevan, ed.) pp. 429-432. Butterworth, London.

Hagem, O. (1970). Anns. Mycol. **8**, 265-286.

Hairston, N.G. (1965). In "Progress in Protozoology", London (1965), pp.114.

Hardie, K. (1983). The effect of removal of extra radical hyphae on water uptake by vesicular arbuscular mycorrhizal plants. New Phytol., **101**, 677-684.

Harley, J.L. (1940). A study of the root system of the beech in woodland soils with special reference to mycorrhizal infection. J. Ecol., **28**, 107-117.

Harley, J.L. and Waid, J.S. (1955). Plant. Soil, **7**, 96-112.

Harley, J.L. (1959). "The Biology of Mycorrhiza", Leonard Hill, London.

Harley, J.L. (1972). The Biology of Mycorrhiza Leonard Hill, London.

Harley, J.L. and Smith, S.E. (1983). Mycorrhizal symbiosis. Academic Press, London.

Harvey, J.V. (1925). J. Elisha Mitchell Scient. Soc., **41**, 151- 164.

Heal, O.W. (1963). In "Soil organisms". (Doeksen, J. and Van der Drit, J. eds). pp. 289-296. North Holland Publishing Co., Amsterdam.

Henderson, M.E.K. (1957). J. Gen. Microbiol. **16**, 686-695.

Henderson, V.E. and Katznelson, H. (1962). Phytopathology, **52**, 423-427.

Henssen, A. (1957). Arch. Mikrobiol., **25**, 63-81.

Hetherington, A. (1933). Arch. Protistenk., **80**, 255-280.

Hessayon, D.G. (1953). Soil Sci., **75**, 395-404.

Hiltner, L. (1904). Arb. dt. LandwGes, **98**, 59-78.

Hofler, K. (1951). Verh. Zool.-bot. Ges. Wien., **92**, 234-241.

Holding, A.J. (1960). J. Appl. Bact., **25**, 515-525.

Horvath, J. (1949). Ann. Ins. Biol. Fer. Hung., **1**, 151-162.

Jackson, R.M. (1958). J. Gen Microbiol **18**, 248-258.

Jackson, R.M. (1960). In "Ecology of Soil Fungi". (Parkinson, D. and Waid, J.S. eds). pp. 168-176 Liverpool University Press.

Jensen, C.N. (1912). Bull. N.Y. (Cornell) Agric. Exp. Stn., **315**, 415-501.

Jensen, H.L. (1934). Proc. Linnean Soc. N.S.W., **59**, 101-117.

Jones, P.C.T. and Mollison, J.E. (1948). J. gen. Microbiol., **2**, 54-69.

Katznelson, H. (1946). Soil Sci., **62**, 343-354.

Katznelson, H.; Ronath, J.W. and Payn, T.M.B. (1954). Nature, Lond. **174**, 1110-1111.

Katznelson, H.; Ronath, J.W. and Payne, T.M.B. (1955). Plant Soil, **7**, 35-48.

Katznelson, H. and Strzelczyk, E. (1961). Can. J. Microbiol. , **7**, 437-446.

Kemper, W.D. and Ameniya, M. (1947). Proc. Soil Sci. Soc. Am., **21**, 657-660.

Kendrick, W.B. and Burges, A. (1962). Nova Hedwigia, **4**, 313-342.

Kerr, A. (1956). Aust. J. Biol. Sci., **9**, 45-52.

Kidston, R. and Lang, W.H. (1921). On the old red sandstone plants showing structure from the Rhynie chart bed Aberdeemshire. Part V. The thallophyte occurring in the peat bed; the succession of the plants through a vertical section of the bed and the conditions of accumulation and preservation of the deposit. Trans. R. Soc., **52**, 855-902.

Kottke, I. and Oberwinkler F. (1986). Root fungus interactions observed on initial stages of mantle formation and hartig net establishment in mycorzhizas of *Amanita muscaria* on *Picea arbies* in pure culture. Can. J.. Bot. **64**, 2348-2354.

Kovada, V.A., Azilevich, N.I. and Rodin, L.E. (1956). In "Takyrs of Western Turkmenistan and ways of their utilization of Agriculture", 22-29. Akad. Naut. U.S.S.R. Moscow (Russian).

Knapp, R. and Lieth, H. (1952). Arch. Mikrobiol. **17**, 292-299.

Krassil' nikov, N.A. (1958). Microoganisms an Higher Plants. Academy of Sciences of the U.S.S.R., Moscow. (Translated by the Israel Program for Scientifc Translation, 1961).

Kubiena, W.L. (1938). "Micropedology". Collegiate Press, Inc., Iowa.

Kuster, E. (1950). Arch. Mikrobiol. **15**, 1-12.

Kuster, E. (1952). Zentbl. Bakt. Paasitkde I. Orig. **158**, 350- 356.

Lenduer, A. (1980). Beitr. Kryptog Flora Schweiz, **3**, 1-180.

Leslie, L.D. (1940 a). Physiol. Zool. **13**, 243-250.

Lesile, L.D. (1940 a). Physio. Zool. **13**, 430-438.

Lewis, I.M. (1928). Cent. Bakteriol., II, **75**, 45-52.

Lewis, D.H. (1973). Concepts in fungal nutrition and the origin of the biotrophy. Biol. Rev. Camb. Philos. Sock. **48**, 261-278.

Leeper, G.W. (1964). "Introduction to Soil Science". 4th Ed. Melbourne. Limitations and Potentials for Biological Nitrogen Fixation in the tropics. (Basic Life Sc. Vol. 10). Ed. Johanna Dohseiner, Robert, H. Burris and Alexander Hillaender. Plenum Press, NY. 1978. Distributed by Brrokhaven Natl. Lai. Associated Univs, Inc, Upton NY.

Linford, M.B. (1940). Phytopathology, **30**, 348-349.

Linford, M.B. (1942). Soil Sci. **53**, 93--103.

Lochhead, A.G. and Nonatt, J.W. (1955). Proc. Soil Sci. Soc. Am. **19**, 48-49.

Lund, J.W.G. (1945-46). New Phytol. **44**, 196-219, **45**, 56-110.

Lund, J.W.G. (1945). New Phytol., **44**, 196-219.

Lund, J.W.G. (1947). New Phytol. **46**, 35-60.

Marks, G.C.and Foster R.C. (1973). Structure, morphogensis and ultra-structure of ectomycorrhizae. In Ectomycorrhizae : Their ecology and Physiology (G.C. Marks and T.T. Kozlowski, eds). Academic Press, New York. pp. 1-41.

Martin, J.P. (1950). Soil Sci. **69**, 215-232.

Marx, D.H. (1969). The influence of ectotrophic mycorrhizal fungi on the resistance of pine roots to pathogenic fungi. Phytopathology, **59**, 153-163.

McCready, C.C. (1952). Studies in the physiology of some ectotrophic mycorrhizas. D.Phil. thesis, Oxford University.

Mc Dougall, B.M. and Rovira, A.D. (1970). Sites of exudation of ^{14}C- labelled compounds from wheat roots. New Phytol. **69**, 999- 1003.

Mishra, R.R. (1965). Seasonal distribution of fungi in four different grass consociations of Varanasi (India). Tropical Ecology **6**, 133-140.

Mishra, R.R. (1966). Studies on ecological factors governing distribution of soil mycoflora. Proc. natn. Acad. Sci. (India) **36**, 205-222.

Mishra, R.R. and Srivastava V.B. (1969). Rhizosphere fungal flora of certain legumes. Ann. Inst. Past. **117**, 17-23.

Mishra, R.R. (1970). Variation in rhizosphere microflora of certain crop plants. Proc. Natn. Acad. Sci. (India) **40**, 195-202.

Mishra, R.R. and Srivastava V.B. (1970). Leaf surface microflora of *Triticum aestivum*. Bull. Torr. Bot. Club. **97**, 364-367.

Mishra, R.R. and Srivastava V.B. (1971). Leaf surface fungi of *Oryza sativa* Linn. Mycopath. Mycol. Appl. **44**, 289-294.

Mishra, R.R. and Srivastava V.B. (1973). Investigations into rhizosphere mycoflora III. Effect of sampling time. Microbiol. Espanola **27**, 115-121.

Mishra, R.R. (1976). Further observations in the phyllosphere microflora of *Triticum aestivum* and *Hordeum vulgare*, Acta Bot. Indica **4**, 111-121.

Mishra, R.R. and Sharma G.D. (1980). Report on root nodule formation in *Eupatorium* species. Curr. Sci. **49**, 444-446.

Mishustin, E.N. (1956). Soils and Fert, **19**, 385-392.

Moore, R. and McClelen C.E. (1983). A morphometric analysis of cellular differentiation in the root cap of *Zea Mays* Aus. J. Bot., **70**, 611-617.

Naim, M.A., A.F. Afifi, and A.a. El-Grindy (1976). Effect of seed and root exudates on some erops as well as their constituents, singly, on the spore germination of some soil micro-mycetes. Ind. Phytopath, <u>29</u>, 412-417.

Nelson, E.B., Chao, W.L. Norton, J.M. Nash, G.T. and. Harman G.E (1986). Attachment of Enterobacter cloacae to hyphae of *lythium ultimum* possible role in the biological control of Pythium preemergence dampin-off. Phytopathology, **76**, 327-335.

Newman, E.I. (1985). The rhizosphere carbon sources and microbial populations. In "Ecological Interactions in Soil" : Plants, Microbes and Animals (A.H.. Fitter, d. Atkinson, D.I. Read and M.B. Usher eds). Blackwell Scientific Publications, Oxford, pp 107-121.

Newman, A.S. and Norman, A.G. (1947). J. Bacteriol, **54**, 37-38.

Nikoljuk, V.F. (1956).. "Soil protozoa and their role in cultivated soils of Uzbekistan", 144 pp. Tashkent.

Nikoljuk, V.F. (1965). In "Progress in Protozoology", London (1965), pp. 118-119. Excerpta Medica Foundation, Amsterdam.

Norman, A.G. (1960 a). Proc. 7th Int. Congr. Soil Sci. II 531- 535.

Nye, P.H. (1979). Soil properties controlling the supply of nutrients to the root surface. In: "The Soil-Root Interface" (J.L. Harley and R.S. Russel, eds). Academic Press, London. pp 39-49.

Orpurt, P.A. and Curtis, J. T. (1957). Ecology, **38**, 628-637.

Osborne, T.G.B. (1909). Lateral roots of *Amyelon radicans* and their mycorrhiza. Ann. Bot. (Cond.) **23**, 603-610.

Oudemans, C.A.J.A. and Koning, C.J. (1902). Archs. Neerl. Sci. Nat. Ser. 2, **7**, 266-298.

Owusu-Bonnoch, E and A. Wild (1979).. Autoradiography of the depletion zone of phosphate around onion roots in the presence of vesicular-arbuscular mycorrhiza. New Phytol. <u>84</u>, 133-140.

Parker, B.C. and Bold, H.C. (1961). Am. J. Bot. **48**, 185-197.

Parker, B.C. and Turner, B.L. (1961).. Evolution, Lacaster, Pa. **15**, 228-238.

Patons, G. (1956). Agrokemia Talajt, **5**, 351-358.

Papavizas, G.C. and Darey, B.C. (1961). Plant and Soil **14**, 215-236.

Parkinson, D., Taylor, G.S. and Pearson, R. (1963). Plant and Soil, **19**, 332-349.

Parkinson, D. and Clarke, J.H. (1964). Plant and Soil. **20**, 166-174.

Parkinson, D. (1965). In "Plant Microbes Relationships", pp. 69-75. Publishing House of the Czechoslovak Academy of Sciences, Prague.

Pearson, V. and Tinker, D.B. (1975). Measurement of Phosphorus fluxes in the external hyphae of endomycorrhizas. In: "Endomycorrhizas" (F.E. Sanders, B. Mosse, and P.B. Tinker, eds). Academic Press, London, pp. 275-287.

Pelczar, M.J., Jr. and Reid, R.D. (1985). "Microbiology", McGraw Hill Book Co., Inc., New York.

Perotti, R. (1926). Proc. Int. Soc. Soil Sci. **2**, 146-161.

Peterson, E.A. (1958). Can J. Microbiol. **4**, 257-265.

Peterson, E.A. (1961). Can. J. Microbiol. **7**, 2-6.

Piccini, D. and Azcon, R. (1987). Effect of phosphate solubilizing bacteria and vesicular arbuscular mycorrhizal fungi on the utilization of Bayovar rock phosphate by alfalfa plants using sand vermiculite medium.

Ponfante - Fasolo, P. (1984). Anatomy and morphology of VA mycorrhizae. In: "VA mycorrhiza" (C. Ll. Powell and D.J. Bagyraj, eds) CRC Press, Boca Raton, F.L. pp. 5-33.

Poschenreider, H. (1930). Zentbl. Bakt. Parasitkde II, **80**, 369- 378.

Raj, J.D., Bagyaraj, J. and Manjunath, A. (1981). Influence of soil inoculation with vesicular - arbuscular mycorrhiza and a phosphate dissolving bacterium on plant growth and 32 p - uptake. Soil Biol. Biochem., **13**, 105-108.

Rayner, M. C. and Levisohn, I. (1940). Production of synthetic mycorrhiza in cultivated cranberry. Nature, Lond., **145**, 461.

Read, D.J. (1983). The biology of mycorrhiza in health and ecosystems with special reference to the nitrogen nutrition of the Ericaceal. In: "Microbial Ecology" (M. W. Lontit and I.A.R. miles, eds). Springer Verlaag, New York, pp 324-328.

Read, D.J. (1983). The biology of mycorrhiza in the Ericales. Can. J. Bot. **61**, 985-1004.

Rehm, H.I. (1960). Zentbl. Bakt. Parasitkde II, **113**, 213-233.

Reid, C.P.P. (1984). Mycorzhizae : a root-soil interface in plant nutrition. In microbial plant Interations, Soil Science Society of America, Madison, W1, pp. 29-50.

Rhodes, L.H. and Gerdermann, J.W. (1978). Hyphal translocation and uptake of sulphur by vesicular - arbuscular mycorzhizae of onion. Soil Biol. Biochem., **10**, 355-360.

Robinson, J.B.D. (1957). J. Agric. Sci, Cambridge, **49**, 100-105.

Rossi, G.M. (1928). Italia Agric. **4**.

Ronatt, J.W. and Katznelson, H. (1960). Nature, Lond. **186**, 659- 660.

Ronatt, J.W and Katznelson, H. (1961). J. Appl. Bact. **24**, 164- 171.

Rovira, A.D. (1953). Aust. Conf. Soil Sci. **1**, 1-7.

Rovira, A.D. (1959). Plant and Soil, **11**, 53-64.

Rovira, A.D. and Bowen, G.E. (1960). Nature, Lond. **185**, 260-261.

Russel, E.W. (1950). "Soil conditions and plant growth". Longmans, Green and Co. London.

Russel, E.W. (1961). "Soil conditions and plant growth". (9th Edition) Longmans, Green and Co. London.

Sanders, F.E. and Tinker, P.B. (1973). Phosphate flow into mycorrzhizal roots. Pestic. Sci., 385-395.

Sartory, A. and Meyer, I. (1948). Compt. Rend. Acad. Sci., **226** 443-445.

Schofield, R.K. (1935). Trans. 3rd Int. Congr. Soil. Sci, Oxford, **2**, 37.

Schroth, M.N. and Snyder, W.C. (1961). Phytopathology, **51**, 389- 393.

Schulze, E.D. (1982). Plant life forms and their carbon, water, and nutrient relation. In: "Physiological Plant Ecology II." (O.L. Lange et al. eds). Springer-Verlar. Berlin, pp 015-676.

Schwabe, G.H. (1963). Pedobiologia, **2**, 132-152.

Sen, A. and Sen, A.N. (1956). J. Indian Soc. Soil Sci, **4**, 215- 220.

Seward, A.C. (1989). Fossil Plants, Vol. I, p. 207. Cambridge University Press, 452 pp.

Shtina, E.A. (1954). Trud. Kirovsk. Sel'sskokhoz. Lust. **10**, 29-69 (Russian).

Shtina, E.A. (1956b). Bot. Zh. SSSR. **41**, 1314-1317. (Russian).

Shtina, E.A. and Young, L. A. (1993). Agrobiologia, **3**, 424- 427.

Singh, B.N. (1945). Br. G. Exp. Path. **26**, 316-325.

Singh, B.N. (1952). Phil. Trans. R. Soc. **236**, 405-461.

Singh, B.N. Mathew, S. and Anand, N. (1956). J. Gen. Microbiol. **12**, 104-111.

Singh, R.N. (1961). "Role of Blue-green Algae in Nitrogen Economy of Indian Agriculture." Indian Coun. Agric. Res., New Delhi.

Smith, N.R. and Humfield, H. (1930). J. Agric. Res. **41**, 97-123.

Smith, S.E., St. John, B.J.. Smith, F.A. and J.D. Nicholas (1985). Activity of glutamine synthetase and glutanate dehydrogenase in *Trifolium subterraneum* L. and *Allium Cepa* L. Effect of mycorrhizal infection and phosphate nutrition. New Phytol., **99**, 211-227.

Sperher, J.I. (1957). Nature, **180**, 994-995.

Stahl, W.H., McQue, B. Mandels, G.R. and Sin, R.G.H. (1949). Arch. Biochem., **30**, 422-432.

Starc, A. (1942). Arch. Mikrobiol; **12**, 329-352.

Starkey, R.L. (1929a). Soil Sci. **27**, 319-334.

Starkey, R.L. (1929b). Soil Sci. **27**, 355-378.

Starkey, R.L. (1929c). Soil Sci. **27**, 433-444.

Starkey, R.L. (1931). Soil Sci. **32**, 37-393.

Stenton, H. (1958). Trans. Br. Mycol. Soc. **41**. 74-80.

Stout, J.D. (1954). J. Protozool. **1**, 211-215.

Strzelczyk, E. (1961). Can. J. Microbiol. **7**, 507-513.

Strugger, S. (1948). Can. J. Res. C **26**, 188-193.

Summerbell, R.C. (1987). The inhibitory effect of *Trichoderma* species and other soil fungi on formation of mycorrzhiza by *Laccaria bicilor* in vitro. New Phytol., **105**, 437-448.

Sylvia, D.M. (1988). Activity of external hyphae of vesicular - arbuscular mycorzhizal fungi. Soil Biol. Biochem., **20**, 39-43.

Szabo, I.; Marton, M. and Szabolcs, I. (1958). Agrokem. Talajt. **7**, 163-175.

Szabo, I.; Marton, M. Szabolcs, I. and Varga, L. (1959). Acta Agron, Hung. **9**, 9-39.

Taylor, C.B. (1936). Proc. R. Soc., B. **119**, 269-295.

Taylor, G.S. (1962). "Studies on Fungi Associated with Roots of certain Grop Plants", Ph.D. thesis; University of Liverpool.

Taylor, G.S. and Parkinson, G.S. (1964). Plant and Soil, **20**, 34-42.

Ternetz, C. (1907). Uber die Assimilation des atmosphari - schen stickstoffes durch philze. Jb. wiss. Bot. **44**, 353-408.

Timonin, M.I. (1948). Proc. Soil Sci. Soc. Am. **11**, 284-292.

Tinker, P.B.H. (1975). Effects of vesicular - arbuscular mycorrzhizas on higher plants. Symp. Soc. Exp. Biol., **29**, 325-349.

Tinker, P.B. and Gildon, A. (1983). Mycorrzhizal fungi and ion uptake. In: "Metals and Micronutrients : uptake and utilization by plants (D.A. Roble and W.S. Pierpoint, eds). Acadamic Press, New York, pp 21-32.

Topping, L.E. (1938). Zent. Bakteriol., II, **98**, 193-201.

Tracey, M.V. (1955). Nature, **175**, 815.

Tribe, H.T. (1957). Trans. Br. Mycol. Soc. **40**, 489-499.

Trofymow, J.A. Coleman, D.C. and Cambardella C. (1987). Rates of rhizodeposition and ammonium depletion in the rhizosphere of axenic oat roots. Plant and Soil, **97**, 333-344.

Vancura, V. and Hanzlikova, A. (1972). Root exudates of plants. IV. Differences in chemical composition of seed and seedlings exudates. Plant and Soil, **36**, 271-282.

Vedkamp, H. (1955). Meded. Landbouw. Wageningen, **55**, 127-174.

Vine, H., Thompson, H.A. and Hardly, F. (1942). Trop. Agric. Trin. <u>19</u>, 215-220.

Volz, P. (1929). Arch. Protistenk, **68**, 349-408.

Volz, P. (1951). Zool. Ja. Abt. Syst. Oekol. **79**, 514-566.

Vrany, J.; Vancura, V. and Macura, J. (1962). Folia Micrbiol., Praha, <u>7</u>, 61-70.

Waid, J.S. (1957). Trans. Br. Mycol. Soc. **40**, 391-406.

Waid, J.S. (1960). In: "The Ecology of Soil Fungi." (D. Parkinson and J.S. Waid, eds.), pp 55-75. Liverpool University Press.

Waksman, S.A. (1916). Scieces, N. S. **44**, 320-322.

Waksman, S.A. (1927). "Principles of Soil Microbiology." Bailliere, Tindall and Cox, London.

Waksman, S.A. and Dishm, R.A. (1931). Soil Sci. **34**, 95-109.

Waksman, S.A. (1959). The Actinomycetes. Vol. 1. Nature occurrence and activities. The Williams and Wilkins Co., Baltimore.

Wainwright, M. (1988). Metabolic diversity of fungi in relation to growth and mineral cycling in soil - a review. Trans. Br. Mycol. Soc. **90**, 159-170.

Warcup, J.H. (1950). Nature, **166**, 117-118.

Warcup, J.H. (1955a). Nature, Lond., **175**, 953.

Warcup, J.H. (1957). Trans. Br. Mycol. Soc. **42**, 45-52.

Warcup, J.H. (1960). In "The Ecology of Soil Fungi." (D. Parkinson and J.S. Waid, eds.)

Webster, J. (1956). J. Ecol. **44**, 517-544.

Webster, J. (1957). J. Ecol. **45**, 1-30.

Weiss, F.E. (1904). Mycorrzhizas from the lower Coal Measures. Ann. Bot. (Lond.) 18, 255-265.

Whips, J.M. and Lynch, J.M. (1985). Energy losses by the plant in rhizodeposition. Ann. Proc. Phytochem. Soc. Eru. **26**, 59-71.

Whipps, J.M. and J.M. Lynch, (1986). The influence of rhizosphere on crop productivity. Adv. Microb. Ecol., **9**, 187- 244.

Wieringa, K.T. (1947). Proc. 4th Intl. Cong. Microbiol, Copenhagen, pp. 482-483.

Willonghby, L.G. (1961). Trans. Br. Mycol. Soc. **44**, 305-332.

Willmer, E.N. (1956). J. Exp. Biol. **33**, 583-603.

Winograasky, S. (1925). Ann. Inst. Pasteur, **39**, 299-354.

Woldendorp, J.W. (1978). The rhizosphere as part of the plant- soil system. In: "Structure and Functioning of Plant Productions" (A.H.J) Freysen and J.W. Woldendorp, eds). Konik. Nederl. Akad. Wetens. Verh., Afdeling Natuurkunde, Tweede, Reeks, Deel 70 North Hooland, New York, pp. 237-267.